In-Vitro Diagnostic Devices

Chao-Min Cheng · Chen-Meng Kuan
Chien-Fu Chen

In-Vitro Diagnostic Devices

Introduction to Current Point-of-Care Diagnostic Devices

Chao-Min Cheng
National Tsing Hua University
Hsinchu
Taiwan

Chien-Fu Chen
National Chung Hsing University
Taichung
Taiwan

Chen-Meng Kuan
National Tsing Hua University
Hsinchu
Taiwan

ISBN 978-3-319-37174-0 ISBN 978-3-319-19737-1 (eBook)
DOI 10.1007/978-3-319-19737-1

Springer Cham Heidelberg New York Dordrecht London

Softcover reprint of the hardcover 1st edition 2016

Printed on acid-free paper

Springer International Publishing AG Switzerland is part of Springer Science+Business Media
(www.springer.com)

Contents

Chapter 1
Introduction to In Vitro Diagnostic Devices

1.1 Overview

Healthcare investment keeps on increasing substantially in recent years [1, 2]. Such investment has also focused on fighting major diseases, enabled by the novel invention of cost-effective and valid drug development for treatment and side effect reduction, along with improved vector control. In addition, the demand for diagnostics that is essential in determining prognosis, identifying disease stages, monitoring treatment, and assessing the spreading as health services has expanded [3].

Molecular-based diagnostics is critical for prevention, identification, and treatment of disease. Current laboratory analyses support correct diagnosis in over 70 % of all diseases and can be used to aid the continuous monitoring of drug therapy [4]. Traditional diagnosis system in central laboratory is therefore a vital component in the clinic and in local general practice. However, classic diagnostic technologies are not completely well suited to meeting the expanded testing requirement because they rely on complicated sample purification and sophisticated instruments which are labor-intensive, timely, and expensive and require of well-trained operators. One of the main challenges for industry is to develop fast, relatively accurate, easy-to-use, and inexpensive devices. For example, microscopy observation requires less infrastructure and is more widely available based on the simplicity and low cost; however, the accuracy is somehow questionable and underutilized (e.g., smear tests for tuberculosis, malaria, and schistosomiasis) [5–7]. As a result, it not only increases the cost and inconvenience of health care but also causes patients to leave the medical system before the diagnostic result is obtained [8]. Faster and more accurate diagnostic tests that require minimum laboratory equipment and operation training play an important role in expanding health care in resource-constrained settings [9, 10].

In addition to the improved efficiency in laboratory diagnostics, there has been a trend toward a more decentralized diagnostics which occurs directly at patients'

C.-M. Cheng et al., *In-Vitro Diagnostic Devices*, DOI 10.1007/978-3-319-19737-1_1

bedside, in outpatient clinics, or at the sites of accidents, so-called point-of-care (POC) systems [11]. The concept of POC testing is mainly for the patient, so short turnaround time, minimum sample preparation and reagent storage and transferring, user-friendly analytical instruments, and digital or visible quantitative or semiquantitative single readout is required [4, 12, 13]. It is clear that on-site or minimum sample preparation and on-chip storage limit the delays that caused by transport and preparation of clinical samples. As a result, shorter turnaround time leads to rapid clinical decision-making and may save fatal consequences. No previous knowledge in sample analysis should be required, so elders can perform the tests at home with minimum training to improve health outcome [14].

The first POC device was urine dipstick test, which was developed in 1957 to measure urinary protein [15]. Glucose meters for diabetic monitoring and lateral-flow devices for pregnancy tests are currently the most widely used devices in POC molecular diagnostics. They are excellent examples of POC tests; however, they are still not applicable if highly sensitive and high-throughput quantitative measurements are required.

In recent decades, some technologies have emerged that fulfill these requirements. Lateral-flow immunoassay (LFIA) devices, for example, which were originally proposed in the 1980s, remain popular largely because of their design simplicity.

Plotz and Singer invented the latex agglutination assay in 1956, from which the technical basis for the LFIA was derived [16]. Plate-based immunoassay was being developed at the same time. The radioimmunoassay was designed by Berson and Yalow in the 1950s [17]. The enzyme immunoassay, which replaced radioisotopes with enzymes, cut down reaction times, and provided higher specificities than a radioimmunoassay, was developed in the 1960s. The fundamental principles of the LFIA continued to be refined through the 1980s and were firmly established during the ensuing years [18, 19]. Since that time, at least another 500 patents have been filed on various aspects of the technology. Several patents have even been formatted by companies such as Becton Dickinson & Co. and Unilever and Carter Wallace.

The chief application driving the early development of solid-phase, rapid-test technology was the human pregnancy test, which was symbolic of continued historical interest in urine testing for medical diagnostic purposes. This particular testing application made great strides in the 1970s, as a result of improvements in antibody generation technologies and significant gains in understanding the biology and detection of human chorionic gonadotropin (hCG), derived largely from the work performed by Vaitukaitis and colleagues [20]. However, to entirely evolve the lateral-flow test, considerable enabling technologies were still required. Many of these technologies, such as nitrocellulose membrane manufacturing, antibody generation, and processing equipment, were developed throughout the 1990s. The purpose of this article is to introduce readers with basic information regarding the LFIA approach.

1.2 Structure

Figure 1.1 displays the key elements of a LFIA. This assay consists of several components, often segmented parts made of different materials. When a test is run, appropriately conditioned sample is added to the proximal end of the strip, the absorbent pad. The treated sample then migrates to the conjugate pad, where an appropriate reagent has been immobilized. The labeled reagent on the conjugate pad can be colloidal gold, or a colored, fluorescent, or paramagnetic latex particle. These specific biological components can be either antigen or antibody depending on the assay format. Next, the sample remobilizes the dried reagent, and particle interaction ensues. Sample and reagent then migrate to the next segment of the strip, the reaction matrix. The reaction matrix is a porous membrane, upon which a final specific biological component has been immobilized. These biological components are usually proteins, either antibody or antigen. They have been bound onto the specific lines of the membrane being used. As the sample and reagent reach this line, they are captured by the applied proteins, and excess liquid moves past this point and is taken up by the absorbent pad. The result is the detectable absence or presence of the test line, read by eyes or by other instruments.

The LFIA may be of two different types: (1) direct (sandwich, Fig. 1.2a) or (2) competitive (inhibition, Fig. 1.2b). Both types can accommodate qualitative, semi-quantitative, and fully quantitative determinations. Direct assay is usually used when testing for larger analytes with multiple antigenic sites, such as hCG, dengue antigen, or human immunodeficiency virus (HIV). A positive result is indicated by the presence of a test line. The conjugated particles also reach and are captured at the control line. The control line typically comprises a species–specific anti-immunoglobulin antibody, specific for the antibody in the conjugate pad. Competitive assay is usually used when testing for small molecules with single antigenic determinants that cannot bind to antibodies on a test line simultaneously. In such cases, a positive result is indicated by the absence of a test line, but a control line may still form.

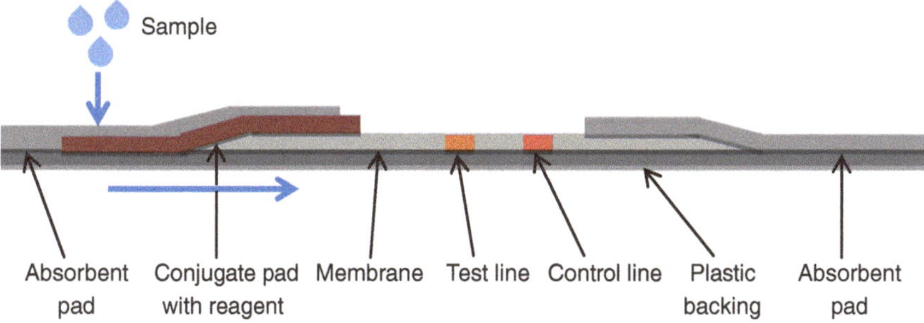

Fig. 1.1 Typical structure of a LFIA strip

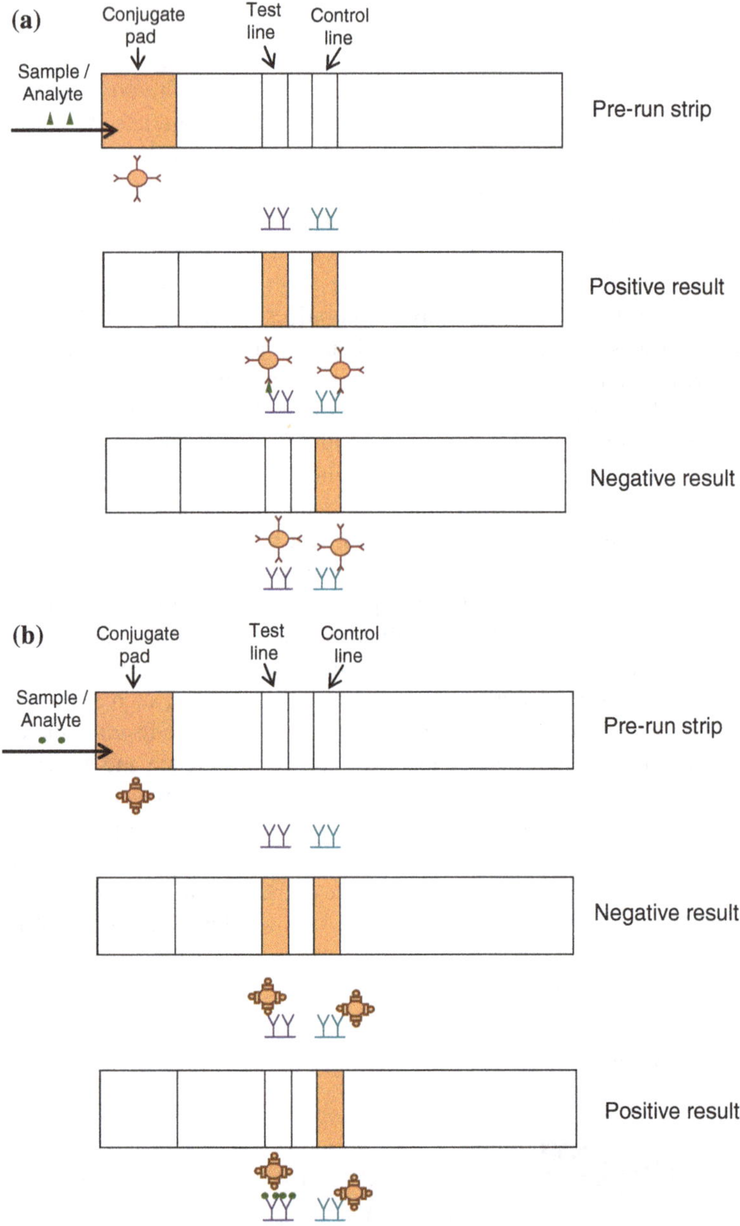

Fig. 1.2 **a** Direct solid-phase immunoassay. **b** Competitive solid-phase immunoassay

1.3 Advantages

LFIAs represent a well-established and very appropriate technology when applied to a wide variety of in vitro diagnostics (IVD) or field-use applications. The advantages of the LFIA are well known:

a. Technology is mature.
b. Manufacture is relatively easy: Equipment and processes are already developed and available.
c. They can be scalable to high-volume production.
d. They can be stored for 12–24 months, often without refrigeration.
e. They are easy to use, requiring minimal operator-dependent steps and interpretation.
f. They can handle small volumes of multiple sample types.
g. They can be integrated with onboard electronics, reader systems, and information systems.
h. They have high sensitivity, specificity, and good stability.
i. Development and approval are relatively low cost and require a short timeline.
j. They are already present and accepted by the market: Minimal education is required for users and regulators.

1.4 Antibody

Although the physical components of the lateral-flow test strip and construction techniques play a major role, the most critical part of the LFIA is the appropriate antibody to provide antigen recognition. If we chose inappropriate antibody, it would not have ability to recognize the target antigen. Much time is spent determining the most suitable antibody for specific assays. Many scientists have spent a great deal of time figuring out the suitable antibodies to fit the assay.

The LFIA is particularly demanding in terms of the mass of the reagent used to drive the antibody and antigen interaction. When an antibody is used in a sandwich-type assay, they are applied at a ratio of 1–3 μg per cm across the width of the nitrocellulose strip, in a line 1 mm wide and with a relatively shallow bed volume of 0.13 mm. This results in an antibody concentration of 10–30 μg per square cm, which is 25–100 times that used in an enzyme-linked immunosorbent assay (ELISA), which can typically require a maximum concentration of 300 ng per square cm [21].

Antibody and antigen affinity also plays an important role in the assay. Consider a typical lateral-flow test strip with antibody immobilized on a test line of 0.5–1.0 mm wide. Antigen flowing up the strip has a flow rate in the range of 0.16–0.66 mm per second, depending on the flow rate of the nitrocellulose membrane selected [22]. Antigen thus spends between 1–6 s on the line where it can interact with the immobilized antibody. Flow speed is actually faster at the initiation of the flow, since the flow rates decrease proportionately to the square of the distance traveled, a steady flow rate is achieved, and the entire nitrocellulose bed volume becomes saturated.

Antibodies applicable for LFIA are available from many commercial sources [19]. Frequently, these antibodies can be obtained for competitive assay, such as hormones, therapeutic drugs, and drugs of abuse. Similarly, suitable antibodies are purchasable for sandwich assay tests to diagnose pregnancy (hCG), infectious disease (HIV, hepatitis B), cardiac markers (troponin C, creatinine kinase-MB, myoglobin), or malignancies (prostate-specific antigen).

1.5 Labels

Some labels have been successfully commercialized and others appear promising. The development of labels for LFIA has matured hand-in-hand with advances in detection methodology and instrumentation. Sensitive assays with fluorescent and luminescent labels have been used in recent years. The ideal labels for lateral-flow strips have the following characteristics:

a. They can be detected by multiple methods on a large and useful dynamic range.
b. When sample and reagent conjugate, their biological and chemical quality and activity are not be changed.
c. The lack of non-specific binding characteristic such as high signal-to-noise ratio under buffer, salt, or detergent conditions.
d. High stability under various temperatures.
e. They are typically available at low cost.
f. The procedure of conjugating is easy and scalable.
g. They are capable of being used for multianalyte detection.

Liposomes can be used as a vehicle for membrane-based assays in vertical and lateral-flow test strips (e.g., test for malarial antigen from Becton Dickinson) [20]. Because of their ability to encapsulate very high concentrations of signal-generating molecules within their cores, liposomes can improve LFIA sensitivity to 2–3 orders. Lipoproteins, glycolipids, and various other lipid-containing compounds can be incorporated directly into the bilayer. In addition, different chemically active groups can be incorporated onto the lipid surface with controlled surface density for covalent coupling to biological or chemical compounds [23].

Colloidal carbon particles can serve as a label in sol particle immunoassays [24]. They have been reported since the 1970s [25]. Their advantages include good stability and high color comparison on a membrane. They are quite easy to conjugate, and a bottle of carbon particles may consequently last for millions of tests.

Colloidal gold has been widely used in immunoassays for large molecules such as for the detection of hormones (pregnancy, fertility), virus (HIV, hepatitis B and C), and bacteria (*Streptococcus suis* serotype 2). It may be the most widely used label today [26]. Determination via colloidal gold-based immunoassay can be completed rapidly in a single step [27]. When an antibody labeled with colloidal gold particles is combined with the corresponding antigen, the colored immunoreactant can be visually detected. This user-friendly format possesses several advantages, including rapid reaction time, long-term stability over a wide range of climates, and low cost. These characteristics make it ideally suited for on-site testing by untrained personnel.

A variety of other labels have been used for specific applications. For instance, a portable fluorescence biosensor with rapid and ultrasensitive response for protein biomarker has been created using quantum dots and a LFIA. The superior signal brightness and high photostability of quantum dots are combined with the promising advantages of a lateral-flow test strip, resulting in high sensitivity, selectivity, and speed for protein detection [28]. Also, more recent reporter up-converting phosphor technology has been applied to DNA (hybridization) assays for the detection of specific nucleic acid sequences. This methodology is sensitive and provides a rapid alternative for more elaborate gel electrophoresis and Southern blotting [29].

1.6 Membranes

While a LFIA test strip may include elegant chemical complexity, the common core of all such tests is the nitrocellulose membrane, which for several reasons is the most significant test component [30–32]. First, it is the surface upon which the critical immune complexes form. Second, it is the surface upon which the signal is detected, either visually or electronically. Third, it has been the most difficult material to manufacture consistently.

One of the key membrane performance parameters is protein binding. It is essential to the function of the membrane in a lateral-flow test strip. The membrane usually adsorbs more than 100 μg of IgG per cm^2. At the concentrations of capture reagents typically applied to the membrane, there is fivefold to tenfold more binding capacity than necessary. Adsorptive capacity decreases with the molecular weight of the protein [33]. To maximize adsorption, antibodies and other proteins should be applied to the membrane in buffers that are preferably free of salt, surfactants, and sugars. The buffer should also be at a low concentration so that crystals dried in the membrane are not of sufficient abundance to occlude the pores.

Another key membrane performance parameter is membrane blocking. Blocking prevents non-specific binding of the detector particle and analyte, but is not absolutely essential to LFIA strips. There are many test strips on the market that do not use a blocking agent; however, blocking agents are required for some tests because of the nature of the particular sample and antibody system [34]. Two blocking agents must be used: one blocking agents dissolves upon addition of the sample and moves along the strip with sample, and the other is applied directly to the membrane by spraying on a fixed amount of blocking solution or dipping the membrane into a reservoir of blocking solution.

The final membrane performance parameter to consider is membrane storage capacity. Storage capacity and condition vary depending on the stage of the test strip manufacturing process. Up until the point that reagent is going to be applied, the membrane can be stored under ambient conditions (15–30 °C, 20–80 % relative humidity). A condensing atmosphere should be avoided, as liquid in the pores can cause redistribution of mobile components, such as the surfactant. When a membrane is being prepared for application of the capture reagents, it should be allowed to equilibrate to the humidity of the dispensing room. Humidity from the air hydrates the surface of the nitrocellulose and improves the absorption of the capture reagent solutions. If possible, assembly of the test strips should take place in a dry room.

1.7 Application

LFIA is well established as a valuable tool in food, medical, environmental, veterinary, agricultural, and industrial diagnostics. Sometimes it is used as a rapid screening tool and backed up by more complex and time-consuming assays. Figure 1.3 lists the market segments in which LFIAs are already in production or are known to be in development.

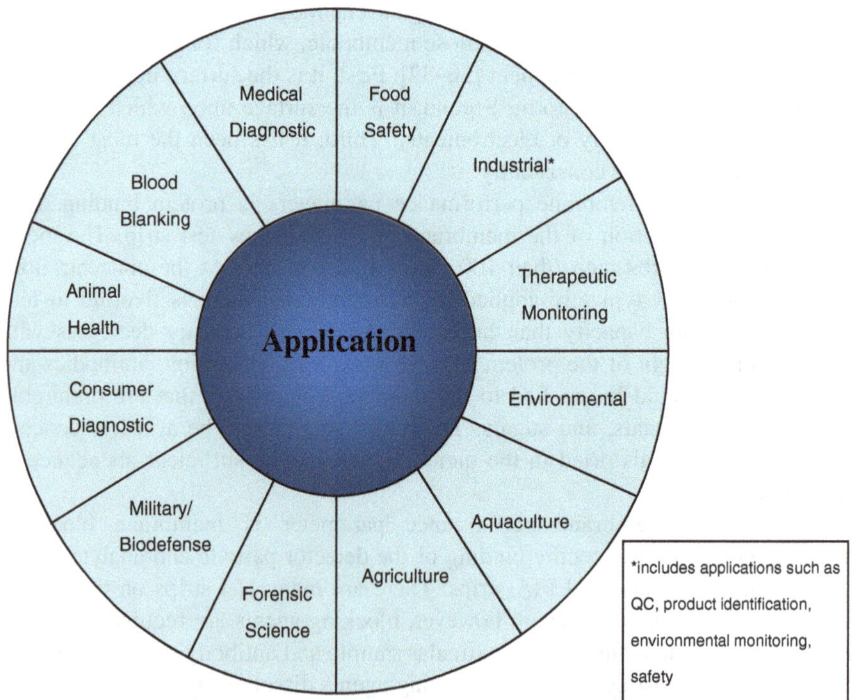

Fig. 1.3 Market segments for LFIA and other point-of-care or field-use technologies

LFIA has well-established formats for POC testing. The first paper-based diabetes dipstick test was created in the 1950s to quantify glucose in urine [35]. Semiquantitative results could be determined by comparing urine-treated test strips to a color-coded chart to determine glucose concentration. Today, commercial urinalysis dipsticks have been widely adapted for a number of analyses. In the 1980s, serological lateral-flow tests started to emerge, particularly for human pregnancy tests. This process was derived from the development of the hCG beta-subunit radioimmunoassay [36, 37].

The majority of these tests come in different sizes, shapes, and configurations. These assays are available without (Fig. 1.4a, b) or with housing units (Fig. 1.4c–f). Nowadays, multiplexing of rapid tests is becoming fairly common as illustrated in Fig. 1.4g–i, which illustrates a lateral-flow format that separates each single lateral-flow test strip into multiple channels. The assay is multiplexed in the sense that a single sample is analyzed simultaneously, but in reality, the test strips are still separate reactions occurring independently of the other reactions [38, 39].

On-chip reagent storage for long-term test and transportation is well developed for IVD. For example, LFIA strips adopt dried gold nanoparticles (AuNPs)-conjugated antibodies regents at conjugation pad for rapid pregnancy, drug abuse, and other diagnostic tests. A plasma fibrinogen assay was implemented on a polymeric micropillar-based LFIA platform by drop-casting bovine thrombin and the surfactant Triton X-100 on the dextran-coated platform [40]. This pillar structure can also be used for an interferon-γ LFIA assay [41].

One of the major application for IVD test is the detection of the metabolites of illegal drugs such as Δ9-tetrahydrocannabinol (THC), amphetamines, benzodiazepines, cocaine, morphine, heroin, opiates, and cannabis in workplace or prison settings. The presence of addictive drugs in the body fluids including blood, urine, sweat, and saliva is monitored to detect and prevent drug abuse, illicit trafficking or driving under the influence of drug (DUID) that is getting more attention worldwide [42, 43]. Furthermore, continuous concern about recreational drug abuse and doping in competitive sports still attracts social attention [44, 45]. The prohibited substances such as strychnine, pervitin, captagon, or Benzedrine are the target molecules for detection.

Oral fluid has been demonstrated as an adequate alternative matrix for drugs identifying and quantifying tests in workplace, clinical treatment, drug rehabilitation center, criminal justice, and DUID settings [46]. The drug tests using oral fluid instead of blood and urine possess various advantages such as inexpensive, rapid, infection risk is lower than for blood sample, and noninvasive of sample collection, which can be easily observed to avoid the need for private facilities and same-sex collectors and decrease adulteration. In addition, oral fluid better reflects recent drug use and reflects free plasma concentrations, providing a better correlation with pharmacodynamic effects.

Liquid chromatography–tandem mass spectrometry (LC-MS/MS) and gas chromatography–tandem mass spectrometry (GC-MS/MS) are the most delegated equipment performing high accurate analysis of multiple compounds in a limited oral fluid volume. However, the complex sample preparation using liquid–liquid

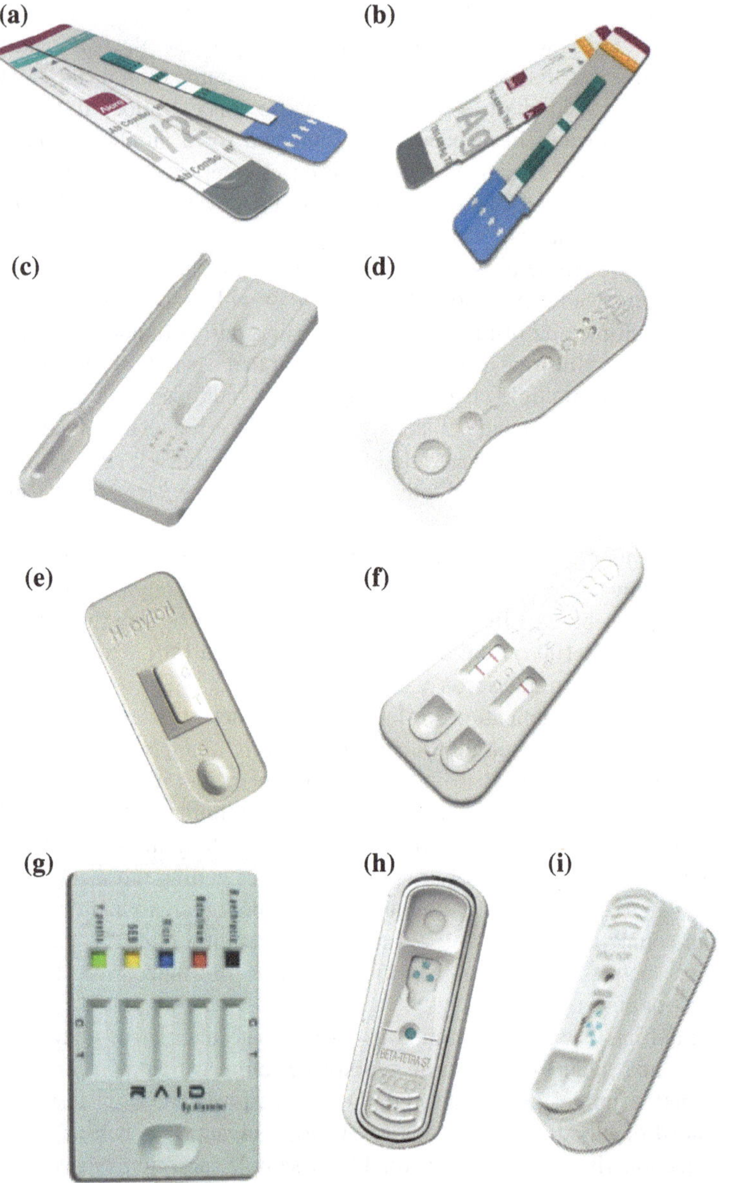

Fig. 1.4 Commercial LFIA tests. **a** Determine™ HIV 1/2 Ag/Ab Combo. © 2013 Alere. All rights reserved. **b** Determine™ TB-LAM Ag test © 2013 Alere. All rights reserved. **c** One Step LH Ovulation Rapid Test © 2010 Accu Plus Medical. All rights reserved. **d** Clearview® Malaria P.f. Test © 2013 Alere. All rights reserved. **e** ICON HP © Beckman Coulter, Inc. All rights reserved. **f** BD™ EZ Flu A + B Test © Becton Dickinson. **g** RAID™ 5 © Alexeter Technologies. **h** SNAPduo™ Beta-Tetra ST Test © 2013 IDEXX Laboratories, Inc. (https://www.idexx.com/small-animal-health/index.html; accessed 10/15/2014). **i** SNAP® Heartworm RT Test © 2012 IDEXX Laboratories, Inc. (https://www.idexx.com/small-animal-health/index.html; accessed 10/15/2014)

extraction or solid-state extraction, time-consuming detection processes, bulky size of equipment, and power sources requirement confined the possibility of on-site tests. There are some commercial portable oral fluid test devices that have been developed and available on market providing satisfactory detection ability to achieve the requirement of detection limit of certain drugs. One of the successful commercial examples for on-site drug test is Oratect. It is a LFIA-based test utilizing AuNPs for colorimetric sensing. In order to collect oral fluidic samples, sample collector is combined in a single device [47].

A number of strategies are available for the detection of nucleic acids in lateral-flow systems [48–50]. The capture of nucleic acids can be performed in an antibody-dependent or antibody-independent way. For example, in an antibody-dependent system, an anti-biotin antibody immobilized on the surface of nitrocellulose is used to capture biotin- and carboxyfluorescein (FAM)-bearing oligonucleotides in RPA amplicons [51]. Binding is subsequently detected using an anti-FAM-colloidal gold conjugate. An antibody-independent alternative utilizes streptavidin as the binding agent. Immobilization of oligonucleotide probes directly onto membranes is also possible using oligonucleotides linked to carrier proteins.

When considering the worldwide market applicability of diagnostics, a socioeconomic division is often applied. Cardiac and other chronic diseases in the expanding middle classes of emerging economies are growing, as are the incidences of previously geographically limited infectious diseases (e.g., malaria, dengue), emerging diseases (e.g., H5N1 influenza), and heretofore well-controlled diseases (e.g., TB in First World countries) in developed countries. At least 30 previously unknown disease agents have been identified since 1973, including HIV, Ebola, hepatitis C, and SARS. In chronic diseases, there remains significant growth, particularly in the areas of inflammation, cardiac markers, and cancer, with a myriad of new labels in development in the search for improved diagnostic and prognostic indicators.

In the past 3–5 years, food safety issues and concerns for public health have led to more stringent legislation in food safety requirements. Legislation has produced increased demand for pathogen and toxin tests in just about every segment of the food production industry. There is a growing demand from food companies for quicker testing to facilitate more rapid release of finished goods and thus reduce inventories. A driver in the demand for rapid and LFIA tests in food production is the adoption of hazard analysis and critical control point (HAACP) regulations that prescribe test procedures throughout the manufacturing process.

1.8 Conclusion

LFIA technology is rapidly being developed. Market needs lead to the improvements in performance and utility and open doors to a vast array of new application areas. With the integration of new reading, labeling, sample handling, and device designs comes a requirement for a new approach to system development and manufacturing. The development of highly sensitive and reproducible/quantitative next-generation point-of-need diagnostic assays requires a different, more multidisciplinary approach than has been the case with standard LFIAs.

References

1. McCoy D, Chand S, Sridhar D (2009) Global health funding: how much, where it comes from and where it goes. Health Policy Plann 24(6):407–417
2. Peeling R, Mabey D (2010) Point-of-care tests for diagnosing infections in the developing world. Clin Microbiol Infect 16(8):1062–1069
3. Nkengasong JN, Nsubuga P, Nwanyanwu O, Gershy-Damet GM, Roscigno G, Bulterys M, Schoub B, DeCock KM, Birx D (2010) Laboratory systems and services are critical in global health time to end the neglect? Am J Clin Pathol 134(3):368–373
4. Luppa PB, Müller C, Schlichtiger A, Schlebusch H (2011) Point-of-care testing (POCT): current techniques and future perspectives. TrAC Trends Anal Chem 30(6):887–898
5. Gray DJ, Ross AG, Li YS, McManus DP (2011) Diagnosis and management of schistosomiasis. BMJ: Br Med J 342
6. Lawn SD, Mwaba P, Bates M, Piatek A, Alexander H, Marais BJ, Cuevas LE, McHugh TD, Zijenah L, Kapata N (2013) Advances in tuberculosis diagnostics: the Xpert MTB/RIF assay and future prospects for a point-of-care test. Lancet Infect Dis 13(4):349–361
7. McNerney R, Daley P (2011) Towards a point-of-care test for active tuberculosis: obstacles and opportunities. Nature Rev Microbiol 9(3):204–213
8. Rosen S, Fox MP (2011) Retention in HIV care between testing and treatment in sub-Saharan Africa: a systematic review. PLoS Med 8(7):e1001056
9. Urdea M, Penny LA, Olmsted SS, Giovanni MY, Kaspar P, Shepherd A, Wilson P, Dahl CA, Buchsbaum S, Moeller G (2006) Requirements for high impact diagnostics in the developing world. Nature 444:73–79
10. Getahun H, Harrington M, O'Brien R, Nunn P (2007) Diagnosis of smear-negative pulmonary tuberculosis in people with HIV infection or AIDS in resource-constrained settings: informing urgent policy changes. The Lancet 369(9578):2042–2049
11. Jani IV, Peter TF (2013) How point-of-care testing could drive innovation in global health. The New Engl J Med 368(24):2319–2324
12. Yager P, Domingo GJ, Gerdes J (2008) Point-of-care diagnostics for global health. Annu Rev Biomed Eng 10:107–144
13. Price C, St John A, Hicks J (2004) Point-ofcare testing. American Association for Clinical Chemistry, Washington DC
14. Price CP, Kricka LJ (2007) Improving healthcare accessibility through point-of-care technologies. Clin Chem 53(9):1665–1675
15. Unold D, Nichols JH (2010) Point-of-care testing: needs, opportunity, and innovation, by Christopher P. Price, Andrew St John, and Larry J. Kricka, eds. Clin Chem 56(12):1893–1894
16. Plotz CM, Singer JM (1956) The latex fixation test: I. application to the serologic diagnosis of rheumatoid arthritis. Am J Med 21(6):893–896
17. Berson SA, Yalow RS (1959) Quantitative aspects of the reaction between insulin and insulin-binding antibody. J Clin Invest 38(11):1996–2016
18. Campbell RL, Wagner DB, O'Connell JP (1987) Solid phase assay with visual readout. 4703017 A, 1987-10-27
19. Rosenstein RW, Bloomster TG (1989) Solid phase assay employing capillary flow. 4855240 A, 1989-08-08
20. Moody A (2002) Rapid diagnostic tests for malaria parasites. Clin Microbiol Rev 15(1):66–78

21. Rowell V (2001) Nunc guide to solid phase. Roskilde, Nunc A/S
22. Rapid Lateral Flow Test Strips (2001) Considerations for product development. Millipore Corporation, Bedford
23. Edwards KA, Baeumner AJ (2006) Analysis of liposomes. Talanta 68(5):1432–1441
24. van Amerongen A, Wichers JH, Berendsen LBJM, Timmermans AJM, Keizer GD, van Doorn AWJ, Bantjes A, van Gelder WMJ (1993) Colloidal carbon particles as a new label for rapid immunochemical test methods—quantitative computer image-analysis of results. J Biotechnol 30(2):185–195
25. Geck P (1971) India-ink immuno-reaction for rapid detection of enteric pathogens. Acta Microbiol Hung 18(3):191–196
26. Chandler J, Gurmin T, Robinson N (2000) The place of gold in rapid tests. IVD Technol 6:37–49
27. Wang S, Zhang C, Wang J, Zhang Y (2005) Development of colloidal gold-based flow-through and lateral-flow immunoassays for the rapid detection of the insecticide carbaryl. Anal Chim Acta 546(2):161–166
28. Li ZH, Wang Y, Wang J, Tang ZW, Pounds JG, Lin YH (2010) Rapid and sensitive detection of protein biomarker using a portable fluorescence biosensor based on quantum dots and a lateral flow test strip. Anal Chem 82(16):7008–7014
29. Corstjens P, Zuiderwijk M, Brink A, Li S, Feindt H, Neidbala RS, Tanke H (2001) Use of up-converting phosphor reporters in lateral-flow assays to detect specific nucleic acid sequences: a rapid, sensitive DNA test to identify human papillomavirus type 16 infection. Clin Chem 47(10):1885–1893
30. Jones KD (1999) Troubleshooting protein binding in nitrocellulose membranes. IVD Technol 5(2):32–41
31. Rapid Lateral Flow Test Strips (2002) Considerations for product development. Millipore Corporation, Bedford
32. Beer HH, Jallerat E, Pflanz K, Klewitz TM (2002) Qualification of cellulos nitrate membranes for lateral-flow assays. IVD Technol 8(1):35–42
33. Mansfield MA (2005) The use of nitrocellulose membranes in lateral-flow assays. In: Wong RC, Tse HY (eds) Forensic science and medicine: drugs of abuse: body fluid testing. Humana Press, Totowa, pp 71–85
34. Weiss A (1999) Concurrent engineering for lateral-flow diagnostics. IVD Technol 5(7):48–57
35. Free AH, Adams EC, Kercher ML, Free HM, Cook MH (1957) Simple specific test for urine glucose. Clin Chem 3(3):163–168
36. Hawkes R, Niday E, Gordon J (1982) A dot-immunobinding assay for monoclonal and other antibodies. Anal Biochem 119(1):142–147
37. Vaitukaitis JL, Braunstein GD, Ross GT (1972) A radioimmunoassay which specifically measures human chorionic gonadotropin in the presence of human luteinizing hormone. Am J Obstet Gynecol 113(6):751–758
38. Eldridge J (2000) Jane's nuclear, biological and chemical Defence 2000–2001. Jane's Information Group Limited
39. Yetisen AK, Akram MS, Lowe CR (2013) Paper-based microfluidic point-of-care diagnostic devices. Lab Chip 13(12):2210–2251
40. Dudek MM, Lindahl TL, Killard AJ (2010) Development of a point of care lateral flow device for measuring human plasma fibrinogen. Anal Chem 82(5):2029–2035
41. Li JJ, Ouellette AL, Giovangrandi L, Cooper DE, Ricco AJ, Kovacs GT (2008) Optical scanner for immunoassays with up-converting phosphorescent labels. IEEE Trans Biomed Eng 55(5):1560–1571
42. Gubala V, Harris LF, Ricco AJ, Tan MX, Williams DE (2011) Point of care diagnostics: status and future. Anal Chem 84(2):487–515
43. Vearrier D, Curtis JA, Greenberg MI (2010) Biological testing for drugs of abuse. Molecular, clinical and environmental toxicology. Springer, Berlin, pp 489–517
44. Deventer K, Roels K, Delbeke F, Van Eenoo P (2011) Prevalence of legal and illegal stimulating agents in sports. Anal Bioanal Chem 401(2):421–432

45. Jelkmann W, Lundby C (2011) Blood doping and its detection. Blood 118(9):2395–2404
46. Gjerde H, Normann PT, Christophersen AS (2010) The prevalence of alcohol and drugs in sampled oral fluid is related to sample volume. J Anal Toxicol 34(7):416–419
47. Wong RC, Tran M, Tung JK (2005) Oral fluid drug tests: effects of adulterants and foodstuffs. Forensic Sci Int 150(2):175–180
48. Seal J, Braven H, Wallace P (2006) Point-of-care nucleic acid lateral-flow tests. IVD Technol 41
49. Dineva MA, Candotti D, Fletcher-Brown F, Allain JP, Lee H (2005) Simultaneous visual detection of multiple viral amplicons by dipstick assay. J Clin Microbiol 43(8):4015–4021
50. O'Farrell B (2007) Sensitive, specific and rapid nucleic acid detection at the point of need using simple, membrane-based assays. Bio World Eur 36–39
51. Piepenburg O, Williams CH, Stemple DL, Armes NA (2006) DNA detection using recombination proteins. PLoS Biol 4(7):e204

Chapter 2
Polymeric-Based In Vitro Diagnostic Devices

2.1 Overview

The concept of translational medicine has begun to change biotechnology as it has encouraged strategic research aimed at transforming multidisciplinary scientific knowledge into real-life healthcare applications [1–3]. Successful applications of compact systems are abundant, including DNA microarrays [4], chip-based polymerase chain reaction (PCR) [5], peptide and oligonucleotide libraries [6], drug screening [7], cell culture [8], and even the concept of living systems on a chip to replace tests using animals [9]. This development demonstrates a growing trend in IVD tests involving versatile and miniaturized lab-on-a-chip (LOC) microsystems that can integrate precise fluid handling, complicated sample processing and signal detection, and readout systems for diseases monitoring, determination of the state of health, or infection detection in order to cure, mitigate, treat, or prevent disease or its sequelae [10]. Advances in bio- and nanotechnology continue to expand the science and relevancy of translational medicine by expanding the scope and capacity of IVD tests [11, 12]. Examples adopting LOC concepts for IVD applications can be easily found in commercial products such as i-STAT (Abbott Point of Care, Princeton, NJ), Dakari CD4 (Dakari Diagnostic, Cambridge, MA), Alere Triage MeterPro (Alere, Waltham, MA), and Piccolo Xpress (Abaxis, Union City, CA), which adopt microelectronic and microfluidic components to create advanced IVD platforms [13].

In addition to conventional analytical materials or microfabrication substrates, such as glass and silicon, polymeric materials have been identified as a good alternative owing to their mechanical flexibility, lightweight, mass fabrication capacity, low cost, and multiple chemical/physical properties based on different grades [12, 14]. Therefore, they gradually become one of the major developing trends for IVD systems. The most common polymeric materials used to fabricate chip-based IVD devices include PDMS, COP, PMMA, PC, and PS [15–17]. A chart summarizing material properties for polymeric IVD device fabrication is provided in Table 2.1.

C.-M. Cheng et al., *In-Vitro Diagnostic Devices*, DOI 10.1007/978-3-319-19737-1_2

Table 2.1 A chart summarizing material properties for polymeric IVD device fabrication

Polymer	Acronym	Tg (°C)	CTE ($10^{-60}C^{-1}$)	Water absorption (%)	Solvent resistance	Acid/base resistance	Biocompatibility	Optical transmissivity	
								Visible	UV
Polydimethylsiloxane	PDMS	−125–122	300–310	0.03	Poor	Good	Excellent	Excellent	Excellent
Cyclo olefin polymer	COP	70–163	60–70	0.01	Good	Good	Excellent	Excellent	Good
Cyclic olefin copolymer	COC	80–180	60–70	0.01	Good	Good	Excellent	Excellent	Good
Poly(methyl methacrylate)	PMMA	100–122	70–150	0.3–0.6	Good	Good	Excellent	Excellent	Good
Polycarbonate	PC	140–148	60–70	0.12–0.34	Good	Good	Excellent	Excellent	Poor
polystyrene	PS	92–106	10–150	0.02–0.15	Poor	Good	Excellent	Excellent	Poor

CTE coefficient of thermal expansion
The variance of these parameters is based on the different grades of polymer

In this chapter, the polymeric material-based IVD systems and the potential IVD systems adopting LOC concepts will be introduced. Rigid polymers such as PMMA, PS, and PC can be used to fabricate less deformable structures. However, their capacity is limited in some lab-on-a-chip applications due to their lower optic transmission and weak organic solvent resistance. Here, we will mainly focus on the properties of two different polymeric materials, PDMS and COP, which are commonly used for new-type IVD prototyping and frequently potential product applications.

2.2 Selection of Polymer Materials

2.2.1 Polydimethylsiloxane

Polydimethylsiloxane (PDMS, Dow Corning Corporation) is probably the most popular polymer due to easy fabrication and bonding for prototyping and testing. PDMS is a commercially available silicone rubber. The physical and chemical properties of PDMS include low glass transition temperature (≈ -125 °C), low loss tangent ($\delta \ll 0.001$), flexibility (shear modulus ~0.25 MPa and Young's modulus ~0.5 MPa), high dielectric strength (~ 14 V μm^{-1}), reasonable temperature variations (thermal expansivity $\alpha \approx 20 \times 10^{-5}$ K^{-1}), wide temperature operation range (from -100 °C up to $+100$ °C), and highly optically transparent from the UV region to the NIR which make it an excellent candidate for optic sensing [18–21]. It is intrinsically hydrophobic with a water contact angle of ~110°, but the surface can be modified by oxygen plasma treatment to become hydrophilic. It can adhere irreversibly, after oxidation, to many different types of substrates [22, 23]. Except at extreme pH values, it has low chemical reactivity. It is characterized by a non-toxic, biocompatible nature with high permeability to O_2 and CO_2 that facilitates cell culturing in lab-on-a-chip fashion [24]. Even some drawbacks such as the stiffness are not as strong as other polymeric materials, surface swollen in organic solvents, and most surface treatment results are often unstable over time possess some limitation [25], other properties are well suited for IVD applications.

2.2.2 Cyclic Olefin Copolymer

As a substrate material for IVD tests, several rigid polymer materials including PC, PMMA, PS, and COP have been considered. Among them, the unique properties of COP, including high resistance to chemicals, high biological compatibility, high transparency in the visible and near-ultraviolet regions of the spectrum, low autofluorescence, low water absorption, low oxygen and moisture permeability, and low gas emissions, make it a strong candidate material for IVD formats [26–28].

COP is highly resistant to acids, alkaline agents, and polar solvents. It is only attacked by nonpolar organic solvents, such as hexane and toluene [29]. This chemical resistance is critical for chip-based sample extraction and separation systems that require multiple washing, loading, elution, and recondition processes [30, 31]. It is also crucial for various bioprocessor, biosensing, monitoring, and screening applications that require multiple chemical reaction processes or a harsh operating environment [32].

The outstanding optical properties of COP allow it to be fabricated as a waveguide and lens material [33]. COP has a high optical transparency over a wide wavelength range from 300 nm to 1200 nm, a large Abbe number, a high refractive index, and a low birefringence, so it can be easily integrated with extra optic systems for sensing. Moreover, in the near-UV region, the transmittance is higher for COP than PMMA, PC, or PS, so it can be used for surface modification by photochemistry synthesis [33, 34]. Another vitally important property of COP is its low autofluorescence, which lowers the background noise when fluorescence detection is employed. The autofluorescence of COP is higher than that of glass or PDMS, but in the same order of magnitude as that of PMMA or PC [33, 35].

The water absorption capacity of COP is about four times less than that of PC and about 10 times less than that of PMMA [36, 37]. This low water absorption provides excellent dimensional stability under a variety of environmental conditions and limits potential solution concentration changes resulting from reagent evaporation and/or absorbance when long processing times are required.

COP is based on ethane and cyclic olefin monomers. Various COP materials are commercially available under brand names including TopasPAS (TOPAS, Florence, KY), APEL (Mitsui Chemicals, Tokyo, Japan), ARTON (Japan Synthetic Rubber, Tokyo, Japan), Zeonex, and Zeonor (ZEON Corporation, Tokyo, Japan). The difference between them is depending on the cyclic monomer and the polymerization process used during synthesis [38, 39]. COP products from Topas and Apel are based on the chain copolymerization of cyclic monomers with ethene, and Arton, Zeonex, and Zeonor are ring-opening metathesis polymerization of cyclic monomers followed by hydrogenation [26, 40]. Even the same brand, different grades with different glass transition temperatures (Tg). Tg increases with a higher cyclic olefin content, so some COP grades possess a higher glass transition temperature than PMMA, PC, and PS [41]. This makes it possible to use certain grades of COP materials in applications exposed to higher temperatures without the risk of deformation and various thermal bonding or micromolding processes [42, 43].

2.3 Fabrication of Polymer Devices

2.3.1 Structure Formation

Versatile fabrication methods are available to structure polymeric materials to form particular geometry for the manufacture of IVD test devices. Laser ablation and micromilling are direct structuring methods suitable for fast prototyping with minimal preparation. Injection molding, also available, is an ideal fabrication process for mass production of commercial available products. Soft lithography, hot embossing, and nanoimprint lithography are more appropriate replication methods for low-cost and laboratory-based production.

2.3.1.1 Soft Lithography

Photolithography continues to be the dominant technology in semiconductor fabrication [44]. As the most important and profitable microfabrication technique, it has contributed to the development of IVD applications including the fabrication of DNA arrays in the late 1980s [45]. However, this technique has a number of limitations, such as an intrinsically expensive fabrication environment and costly equipment, surface modification difficulties, and obstacles to plain surface morphology manipulation that decrease its application in biomedically relevant research fields.

Soft lithography is a technique based on microstructure formation, molding, and embossing to obtain a reverse elastomeric stamp (Fig. 2.1) [46, 47]. These techniques were developed as an alternative to photolithography. No specific laboratory environment is required, and the process does not involve expensive equipment. Soft lithography is a non-photolithographic strategy based on self-assembly and replica molding for carrying out micro- and nanofabrication. It can continuously create large three-dimensional features that can be used in an ordinary laboratory without the need for clean room facilities. In its initial steps, soft lithography relies on the use of photolithography to generate a master used for replication. Once the master is fabricated, the fabrication tasks can be performed outside of a clean room via printing or molding procedures. A large number of patterning techniques such as replica molding [48], microtransfer molding [49], solvent-assisted molding [50], micromolding in capillaries [47], phase-shifting edge

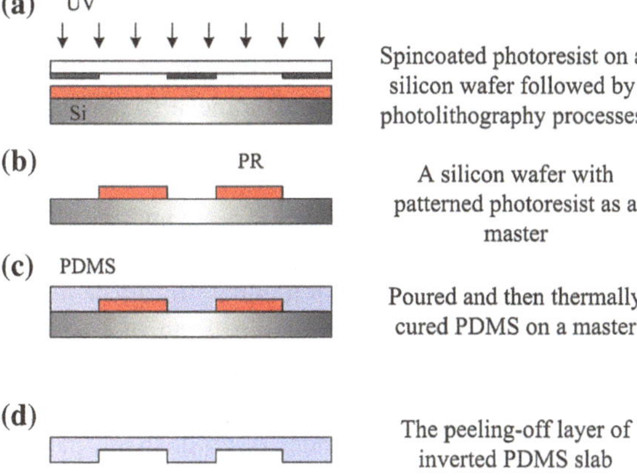

Fig. 2.1 The fabrication of PDMS slab using soft lithography. **a, b** Master is first formed by spincoated photoresist on a silicon wafer followed by photolithography processes. **c** PDMS mixture is then poured on the master and cured thermally. **d** The peeling-off layer of PDMS slab has invert microstructures to the master

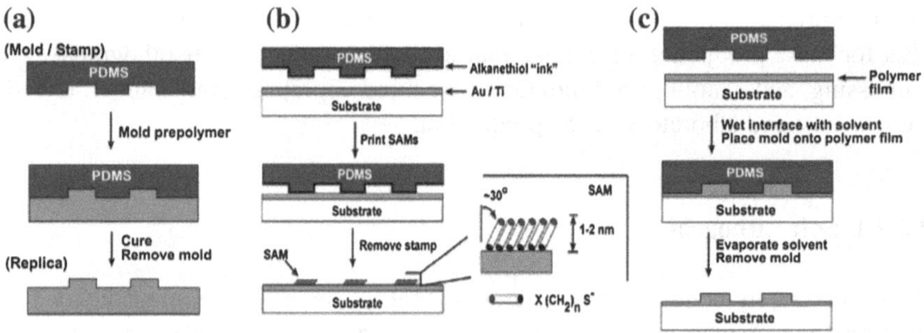

Fig. 2.2 Schematic illustration depicting the procedure for **a** replica molding (RM), **b** microcontact printing (μCP), and **c** solvent-assisted micromolding (SAMIN) [58]

lithography [51], decal film transfer lithography [52], nanotransfer printing [53], microcontact printing [54], nanoskiving [55], and dip-pen nanolithography [56, 57] have been developed. Soft lithography is a cost-effective option that allows for the use of adjustable surface chemistries, requires a minimal laboratory environment, and is highly compatible with biological applications including cell biology, microfluidics, and a variety of lab-on-a-chip systems.

In the following section, three commonly used soft lithographic techniques are introduced, including replica molding, microcontact printing, and solvent-assisted micromolding (Fig. 2.2) [58].

Replica molding is a process that transfers a pattern from a rigid or elastomeric master mold into another material via solidification of liquid poured into the mold. This new method for fabrication of PDMS-based IVD devices for prototyping [22] has proven to be particularly suitable for numerous biomedical device applications [59–62]. Because the production of PDMS microstructures is simple, it can be readily used to make prototype devices and full-function integrated systems [10, 63]. It is also an attractive process for nanofabrication of devices with lateral dimensions smaller than 100 nm [64]. This technique has also been used to micropattern biocompatible polymers including epoxies, polyurethanes, polyethylene glycol (PEG), agar, and agarose for isolating and culturing bacterial cells.

Microcontact printing (μCP) is a large-area (>cm^2) patterning technique used on functional organic surfaces. The process is similar to using a common stamp to transfer ink from an ink pad to a piece of paper. In this process, a mold is stained with a chosen material, e.g., small biomolecules, proteins, polyelectrolytes, or suspensions of cells, and this material is transferred to the substrate surface when contact is made between the substrate and the protruding features of the stamping mold. μCP has been successfully used to print precise patterns of axon guidance molecules as a cell growing template for growing chick retinal ganglion cell axons [65]. It also allows for the engineering of surface properties via molecular-level detailed adoption of the self-assembled monolayer (SAM) technique on the substrate when PDMS stamps stained with alkanethiols ($SH\text{-}(CH_2)_n\text{-}X$) are used to microcontact print on surfaces of gold, silver, palladium, platinum, or other metals

[66]. Formation of alkanethiolate SAMs include thin-film, physical vapor deposition on silicon, mica, glass, or plastic materials [67]. Patterned SAMs are valuable for studying the role of spatial signaling in biosensing and cell biology by controlling the molecular structure of a surface in contact with cells and proteins on different platforms [56].

Solvent-assisted micromolding (SAMIN) is similar to replica molding and is based on molding or embossing with an elastomeric stamp. In this procedure, an elastomer mold is wetted with a solvent before the conformal contact is made between the elastomer mold and the substrate. The liquid solvent fills the recessed regions on the elastomer mold contact surface, which minimizes the area of the liquid/vapor interface and maximizes the solid/liquid interface. As a result, nanoscale structures can be produced in various soft materials over large areas (>cm^2 with 100 nm features). This process can also be combined with selective etching and liftoff processes to transfer into metals that can then be used as substrates for various biomedical sensing platforms such as electrochemistry, [57] surface plasmon resonance [68], optical diffraction [69], and surface-enhanced Raman scattering [70].

2.3.1.2 Injection Molding

Injection molding is a scalable strategy for manufacturing thermoplastic materials with features on the order of micrometers and above. It is one of the most common techniques for the fabrication of polymeric products, since it is highly adaptable for mass production [71, 72]. The process involves initially feeding polymeric pellets into an injection molding machine hopper and then applying high temperature to melt the pellets before the mass is injected into a mold and high pressure is applied. This constant packing pressure is applied for a brief time before the polymeric material and mold are cooled and the manufactured piece is demolded.

The quality and fidelity of the replicated structures depends to a great extent on the master and the fabrication processes. The high pressure and temperature ranges used when molding limits the use of silicon, glass, resists, and other polymers as mold material, so metal materials are commonly chosen [73]. Those molds with micrometer resolution for LOC applications are usually fabricated using standard photolithography techniques followed by electroplating to prolong the mold's lifetime [74–76]. A master mold can withstand more than 200 cycles without severe deformation [77].

Many parameters, including injection speed, mold temperature, and structure geometry in the injection molding process, have a direct effect on polymeric replica quality and usability. Because injection masters may incur structural disruption after a number of cycles, surface replication patterns are influenced not only by pressure distribution inside the mold but also by internal structural deformation [78]. This can often be resolved, and master cavities can be filled more effectively by using higher injection temperatures that produce better flow behavior [79].

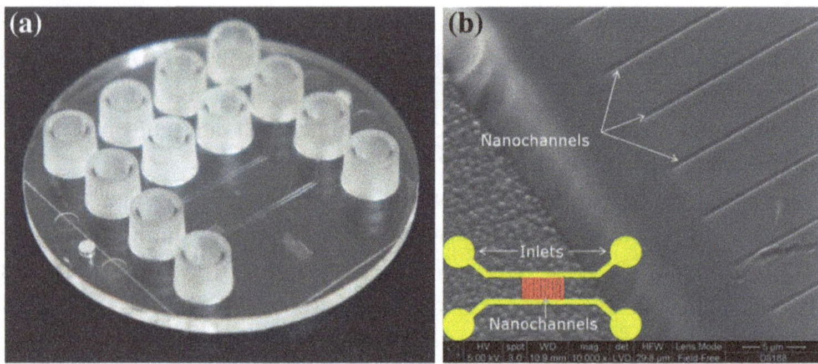

Fig. 2.3 **a** Image of injection molded polymer chip and **b** scanning electron micrograph (SEM image of the nanochannels on the injection molded portion of the device [80]

Injection molding has been used to produce LOC systems for on-chip liquid chromatography and DNA bar coding systems [43, 80]. However, this process can be challenging because of the large aspect ratios required, low surface roughness, high flow pattern resolution, and stresses experienced by the replica. Replication of nanostructures into polymer surfaces has been successfully achieved and had received a good deal of attention (Fig. 2.3) [80, 81]. The main limitation of this process lies in the grain size of the master, which limits its ability to successfully replicate smaller structures.

2.3.1.3 Hot Embossing

Hot embossing is a technique that employs a polymeric sheet pressed onto a microstructured mold or wafer heated above the glass transition temperature (Tg), followed by demolding at approximately Tg −50 °C [42, 82, 83]. Figure 2.4 depicts a schematic of the hot-embossing process [84]. The applied pressure to hold the template and sheet, pressed time, and operation temperature are the most important factors influencing the end quality of the embossed structures on the chip surface. An anti-sticking layer is usually used to provide good fidelity for large-area embossed structures. The chip or wafer format polymeric substrate can be acquired commercially or formed by injection molding or by heating polymer pellets above Tg [85].

Hot embossing is generally done using lower fabrication temperature ranges and lower pressure compared to injection molding. Based on the lower pressure and temperature used, silicon, copper, nickel, stainless steel, and even polymer masters have been used as template material [85, 86]. Among them, SU-8 photoresist has also been used as a master material through standard photolithography for embossing polymeric materials under conditions similar to the ones used with metal templates, and an aluminum coating can be applied to facilitate substrate demolding from the template [87]. To fabricate high aspect ratio structures, an

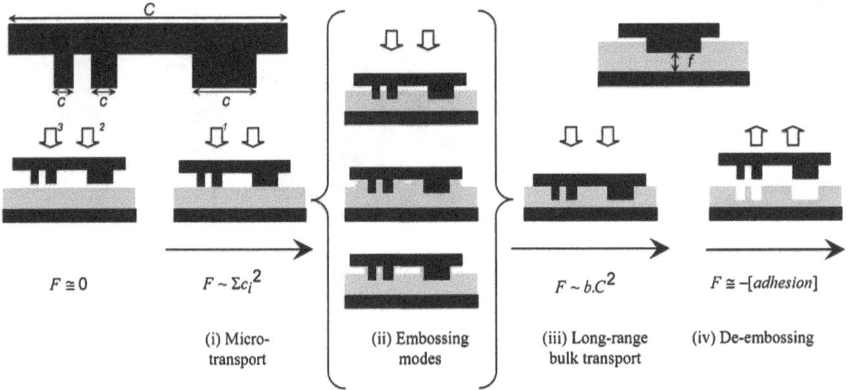

Fig. 2.4 The schematic hot embossing process. The total force (F) required to emboss a thermoplastic polymer depends on the polymer's viscosity, contact area of the stamp features with the polymer (c), surface area of the entire stamp (C) and temperature [84]

anodic aluminum oxidation nanostructured membrane and cylindrical pillars can be used [88, 89].

The distance that the substrate must flow toward the template in the hot-embossing process is smaller than that for high-temperature injection molding, which leads to reduced stress and shrinkage effects during operation [90]. Lower mold temperatures and cooling rates lead to production of more fragile structures with higher aspect ratios than those achieved with injection molding.

2.3.1.4 Nanoimprint Lithography

Nanoimprint lithography (NIL) is a non-conventional lithographic technique for low-cost, high-resolution patterning of polymer nanostructures. During the NIL process, which is performed in a vacuum chamber with two parallel templates pressed together, the substrate is first heated to a temperature above the Tg and then, the templates are pressed against the substrate at a pressure for a period of time. After imprinting, the substrate is cooled down, under constant pressure, to a temperature lower than Tg to avoid deformations and demolding is performed at room temperature. In contrast to traditional lithography, which uses photons or electrons to modify the chemical and physical properties of the resist to achieve high definition patterns, NIL relies on direct mechanical deformation of the molded material to achieve high-resolution patterning (Fig. 2.5) [91]. In hot embossing, only a small portion of the substrate surface is embossed into the template mold; however, because NIL uses most or all of the polymeric film thickness for pattern production, the residual polymer layer left after substrate imprinting is quite thin and may be nonexistent [92]. O_2 plasma in a standard reactive ion etcher system is usually used to etch the remains of the residual layer. The etching

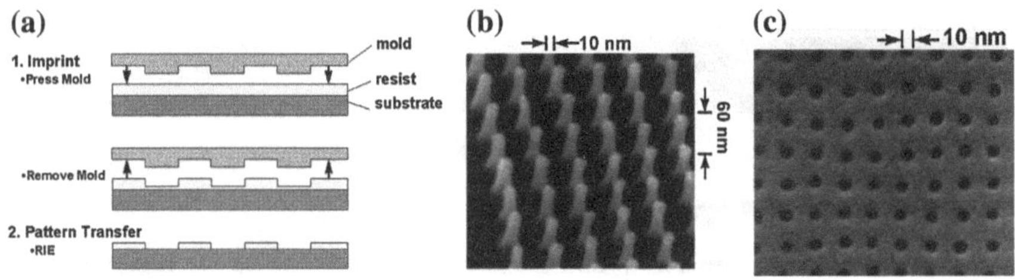

Fig. 2.5 **a** Schematic of the NIL process. SEM images of **b** a fabricated mold with a 10-nm-diameter array, and **c** hole arrays imprinted in poly(methyl methacrylate) by using mold in **b** [91]

process produces heat to the sample; somehow, it would cause polymer to reflow and affect the morphology [92].

The quality of the imprinted replica structure depends, primarily, on the master used. Due to the low pressures and the moderate operation temperatures used, silicon is a good template material [93]. When silicon is used, the pattern is transferred to a silicon wafer using photolithography, or electron beam lithography for smaller structures, followed by deep reactive ion etching (DRIE) to create high aspect ratio structures. The sidewall roughness caused by photolithography or the etching process can be decreased after thermally oxidizing the silicon template [93]. As with other demolding processes, an anti-sticking layer is recommended for small features with high fidelity. NIL can also be carried out by spinning a liquid polymeric solution onto a silicon, silicon dioxide, or glass wafer [94]. If polymeric pellets are used, dissolvation in a proper nonpolar solvent is required. The substrates are baked after spin coating to remove the solvent in which the polymeric materials were diluted.

2.3.1.5 Direct Machining

Direct structuring laser ablation and micromilling are two techniques that can be used for rapid prototyping of polymeric materials without the need for templates or molds.

In laser ablation, the interaction of a high-intensity laser beam with the polymeric material causes the latter to evaporate at the laser focal point [95]. The deeper structures, cluster of redeposited material that is formed close to the ablated holes, are then removed by ultrasonic cleaner. Laser ablation has primarily been used on PMMA [96]. A disadvantage of the ablation technique is that it can lead to polymer surface property change in comparison with a polymer's bulk properties [97]. This is difficult to control and critically important in biosensing applications where the surface chemistry plays a major role in biomolecular receptor binding events that are directly relative to sensing performance.

Mechanical milling of polymeric structures is also a commonly used method for rapid prototyping [98, 99]. Milling consists of patterning a substrate with a

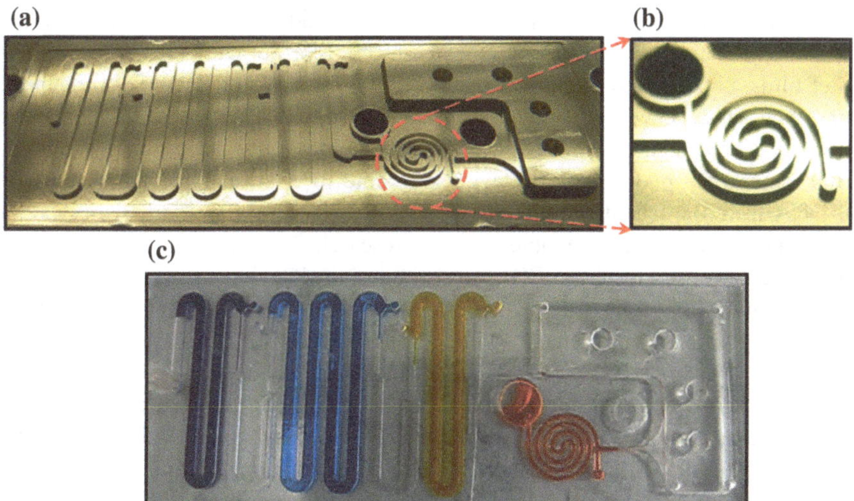

Fig. 2.6 **a**, **b** Image of the milled aluminum master using CNC milling machine and **c** the replicated polymeric device for the sample-to-answer immunoassay chip [102]

rotating end mill or drill controlled by a computer numerical control (CNC) milling machine [100]. The dimensions of milled microstructures are mainly defined by the diameter of the mills themselves. The feed speed of the mills, the properties of polymeric materials, and the system stability affect both the resolution and the milled surface roughness [101]. Multiple components, from macrostructures to microstructures, can be fabricated on the same chip simply by changing the mill size and milling pathway. In addition, aluminum or brass can be milled using a CNC milling machine as a master and then, hot embossing may be performed to fabricate highly durable polymeric IVD chips (Fig. 2.6). A sample-to-answer immunoassay chip, for example, can be continuously replicated for the detection of thyroid-stimulating hormone [102]. However, with mechanical milling, it is not easy to obtain complex, three-dimensional high aspect ratio structures at the micrometric scale.

2.3.1.6 Laser-Printed Microfluidic Devices

As described in a previously reported review article, toner- and paper-based devices comprise the latest generation of disposable microfluidic platforms [103]. Toner-based devices are fabricated by laser printing, a method proposed nearly ten years ago by do Lago et al [104]. Such devices, producible in a matter of minutes, are most often printed on a polyester film surface, the end result being the generation of polyester-toner (PT) devices. These devices, often called PT chips, have demonstrated potential for DNA-based, colorimetric bioassay, and immunoassay investigations.

Duarte et al. were able to integrate dynamic, solid-phase DNA extraction and PCR amplification steps using a PT device [105]. Starting with two separate polyester films, both printed with toner layers and laser cut with microfluidic channels, they then sandwiched and laminated these layers with base and cover polyester films that were manufactured with holes to provide access to laser-cut channels. This remarkable multilayer PT device was able to recover approximately 65 % of DNA from 0.6 μL of sample blood as well as successfully amplify the 520 bp fragment of the λ-phage genome. In other research, Duarte et al. [106] leveraged the low electroosmotic flow (EOF) magnitude in a PT chip to successfully carry out as many as five consecutive DNA fragment separations on their PT electrophoresis device without replacement of the porous matrix.

de Souza et al. [107] developed toner-based microfluidic devices capable of performing clinical diagnostics via capillary action and colorimetric detection. The authors demonstrated that they could integrate detection zones with microfluidic channels for rapid sample distribution via capillary action. Adding an intermediary polyester film encouraged spontaneous fluidic transport, as it increased channel depth as well as the aspect ratio of the channel. Colorimetric assays of artificial human serum samples for glucose, protein, and cholesterol have been successfully performed using a desktop scanner. The LOD values found for glucose (0.3 mg/mL), protein (8 mg/mL), and cholesterol (0.2 mg/mL) were associated with the dynamic range reported by the authors and indicate that this platform is suitable for clinical assays.

Two different groups have successfully demonstrated the use of toner-based platforms to perform immunoassays. First, Oliveira et al. [108] described the quick and simple fabrication of toner-based 96-microzone plates to detect dengue virus. They created detection zones (wells) by printing a hydrophobic toner layer (*ca.* 5 μm thick) that acted as a barrier to confine small sample volumes, and used a cell-phone camera to record and examine their colorimetric results. Dengue virus was detected in human serum samples from infected patients based on capture ELISA of immunoglobulin M (IgM) antibody, a specific marker related to the primary infection of dengue. In the second, and more recent toner-based platform study, Kim et al. reported the use of PT microchips to perform immunoassays capable of detecting C-reactive protein (CRP), a highly conserved plasma protein related to the inflammatory state [109]. By immobilizing this protein on the surface of silica microbeads, placing them in the PT microchannel, and adding a detection antibody with a fluorescent tag to the functionalized surface, a cleavable protein–target complex could be created. Upon cleaving, the fluorescent tags could be analyzed via microchip electrophoresis. The time needed for the complete analysis to be carried out on a PT microchip was less than 35 min. The dynamic range of the CRP in 10-fold diluted serum was 0.3–100 mg/L, and the LOD achieved was 0.3 mg/L, which demonstrated the possibility for quantitative analysis of CRP in serum in clinical trials.

2.3.2 Device Sealing

2.3.2.1 Adhesive Bonding

The fabrication of lab-on-a-chip systems involves a bonding step by which microchannels and compartments are sealed including steps that employ screws or fasteners to close. Gluing, by applying a bonding agent to the interfaces of two separated polymer parts, is the most common bonding technique to seal objects. A thin layer of PDMS pre-polymer, for example, can be spin coated onto a glass slide and then transferred onto the surface of patterned substrate via direct contact. This coated substrate is brought into contact with a flat plate, and the two structures are permanently bonded to form a sealed fluidic system after thermocuring the pre-polymer. The PDMS exists only at the contact area of the two surfaces, and only a negligible portion is exposed to the microfluidic channel. This method has been demonstrated by bonding microchannels of two representative PDMS substrates and two representative glass sheets [110]. The drawback to this process is the possibility of clogging when the pressed glue layer flows into microstructures.

Lamination of polymer layers has been used for the fabrication of IVD devices based on its ease of operation and low cost [15, 111]. Microchannels can be formed directly in a pressure-sensitive adhesive (PSA) or thermobond adhesive (TBA) film using a desktop cutting instrument such as a CO_2 laser, die cutter, or vinyl cutter followed by application of two plastic sheets pressed onto both PSA surfaces. The fluidic channel geometry and thickness are defined by the patterns on the PSA and the thickness of PSA, respectively [112]. One of the primary advantages of using PSA is that it can bond to different polymeric materials easily with minimal physical or chemical manipulation [113, 114]. Three-dimensional microstructures can be formed by laminating and aligning multiple plastic layers together using PSA films. As a result, a fully integrated immunoassay card enabling quantitative assay with an IVD cartridge that combines dry reagent storage, conjugate pad, and microchannels has been realized (Fig. 2.7) [114]. This work was a further part of a larger effort to develop a microfluidic point-of-care system, the DxBox, for sample-to-result differential diagnosis of infections that present with high rapid-onset fever [115]. In this application, dry reagents can be reconstituted prior to use, which eliminates the need for refrigerated storage. However, while glue-based and PSA adhesion is convenient and rapid, the significant limitation, structure resolution, is only several hundred microns and resulting surface roughness can be problematic.

2.3.2.2 Thermal Bonding

Direct thermal bonding of polymeric materials is frequently adopted to enclose polymer devices [116–118]. The mechanism is based on the idea that heat increases the diffusion of the polymer chains between the mating parts of

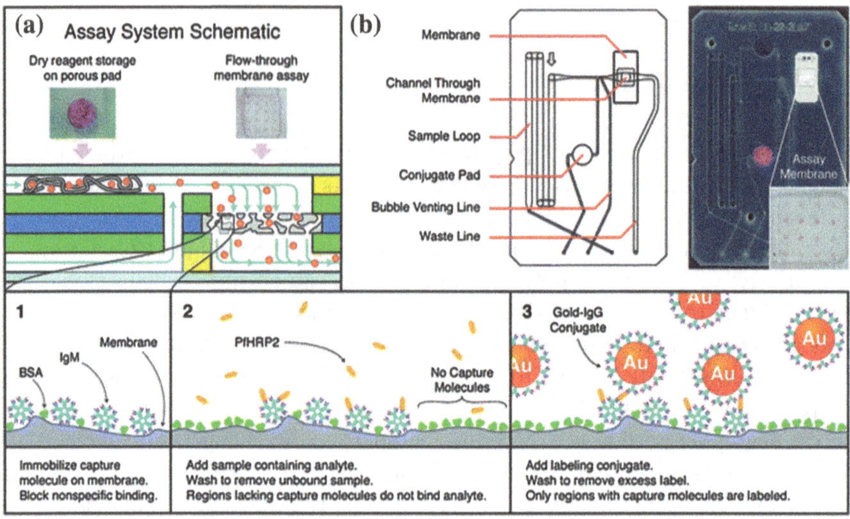

Fig. 2.7 **a** Cross section and close-up schematic of the flow-through membrane assay format. **b** Design and image of assembled, 10-layer assay card. The card is pictured before use, with the red gold–antibody conjugate present in the pad. The *inset* image shows the pattern of capture regions visible on the membrane after completion of the assay [114]

contacted surfaces. Polymer can be thermally bonded by heating to the Tg of the polymer and pressing them together for a period of time to strengthen the diffusion. The bonded substrates are then cooled to Tg −30 °C to avoid the reflow of polymer, and then, applied force is removed. Bond strength evaluation has demonstrated that temperature is the most critical parameter in lamination. If the bonding temperature is too high, it leads to channel deformation and collapse, whereas low temperature is insufficient for complete bonding. In addition to uniform pressure distribution, applied pressure must be optimized to ensure full contact at the bonding interface and avoid collapse. One of the major limitations of thermal bonding is relatively low bond strength due to the low surface energy of thermoplastics.

Plasma activation, such as UV/ozone and O_2 plasma, has been shown to improve the bond strength of polymer substrates [119, 120]. This advanced surface treatment can increase the surface energy and enhance the interdiffusion of polymer chains between mating surfaces during bonding. Plasma activation is usually used in combination with thermal bonding to lower the bonding temperature and avoid deformation.

2.3.2.3 Solvent Bonding

Solvent bonding is another bonding technique especially suited for polymer IVD devices that require high bond strengths [121]. The bonding process includes exposing one or both bonding surfaces to solvent vapor to absorb solvent and then bringing the mating surfaces into contact under pressure for a period of time. The bonding mechanism is similar to thermal bonding except that a solvent is used to promote entanglement of the polymer chains rather than just heat; this creates more mobile interdiffusion across the bonding interfaces. Solvent vapor exposure time and bonding pressure are the two key factors that affect bonding results. If the polymer surface processes solvent vapor exposure for too long time, the surface would uptake too much solvent and cause swelling. If the solvent exposure time is not long enough, no double bond or interdiffusion will occur. Pressure is vitally important because it ensures complete contact between bonding surfaces and eliminates the possibility of voids or deformation. If created under optimized conditions, sealed polymer sheets can withstand pressures greater than 20 MPa without delamination [121]. Notably, a novel solvent vapor treatment has been used to irreversibly bond PMMA and COC chips while simultaneously reducing the channel surface roughness and yielding optical grade, which is less than 15 nm surface roughness on the channel walls (Fig. 2.8) [122]. In conclusion, reflow of polymer with low-cost prototyping tools has been demonstrated to create uniform bonding and produce optical quality surfaces when executed with thought and careful preparation.

Bonding strength in these devices can be further improved by exposing one of the bonding surfaces to solvent vapor, bringing the mating surfaces into contact, and then exposing the stack to UV light. This promotes a higher polymer chain mobility at the bonding intersurfaces. As a result, an improved burst pressure of 34.6 MPa, regarded as the highest in pressure resistance at the time of publication, can be achieved [43].

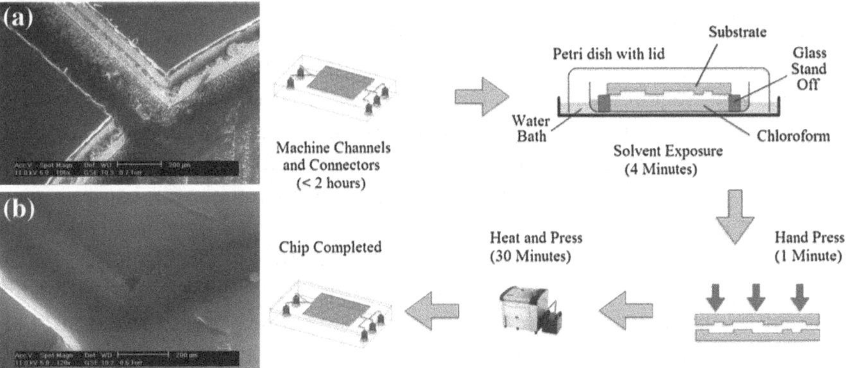

Fig. 2.8 **a** SEM images of milled PMMA microchannel and **b** after 4-min chloroform solvent vapor and 30 min 60 °C heat cycle. **c** Schematic of the solvent bonding processes [122]

2.3.2.4 Welding

Localized bonding can be done by inducing heat and softening the interfaces between mating surfaces using ultrasonic energy [123]. Microwave energy, alternatively, has been used in a process that deposits metallic films on mating surfaces before heating [124, 125]. Polymer sheets can also be bonded using infrared laser welding, which is based on bonding substrates at particular infrared wavelengths. In laser welding, an opaque surface is used for energy absorption and localized heat generation for enhanced bonding. The advantage of welding is that the bonding temperature is below Tg because very thin coating layers are used as an absorbing layer, which eliminates structural damage caused by polymer reflow. Bond strength is adjustable according to bonding temperature. Higher bonding temperature results in a stronger bonding strength.

2.3.3 World-to-Chip Interface

Microfludics technology has been used for point-of-care platforms that integrate analytical processes and functions into a single compact system. Off-chip components such as pumps, valves, and other functional units for sample processing are desirable for on-chip analysis. As a result, fluidic interfaces capable of seamlessly transferring analytes, reagents, and other solutions between external control systems and analytical chip are necessary (Fig. 2.9) [126–131].

In addition to being inexpensive and easy to use, an interconnect should be avoid dead volume appeal within the flow path. Dead volume occurs readily at the flow interconnects, and it results in analyte dispersion or even sample loss. Dead volume can trap an air bubble and introduce non-compliance in a fluidic system resulting in fluid transport delays and potentially leading to channel clogging.

Due to the wide variety of substrate materials, different fabrication processes, and functional requirements of IVD systems, there is no standard world-to-chip interface [132–136]. Low dead volume fluidic interfaces with glass and silicon-based chips rely on modifications to the variety of manifold assemblies [137–139].

Polymer microfluidic chips fabricated from rigid thermoplastics can use interconnecting capillaries partially inserted into the imprinted channels and seamlessly secured by thermal deformation of the substrate [140]. Efficient fluidic interconnectors to elastomer polymer-based devices such as PDMS are readily achieved by directly forming a hole in the elastomer via in situ molding [141] or post-cure punching [131]. Tubing can be directly inserted into polymeric chip (Fig. 2.9a) or connected to the polymer chip using a stainless steel tubing that enhances the stiffness and robustness (Fig. 2.9b) [127, 128, 130, 142]. Another interesting removable connector design is to use a magnetic connector made of

Fig. 2.9 **a** Tubing is directly inserted into polymeric chip or **b** connected to polymer chip using a stainless steel tubing first to connect to external pumping systems. **c** A removable connector design is to use magnetic connector made of a ring magnet with a hole that accommodates tubing or a needle and second magnet placed on the backside of the chip to prevent leakage [129, 130, and 142]

a ring magnet with a hole that accommodates tubing or a needle (Fig. 2.9c). The tubing/needle is fixed to the magnet with epoxy, and a gasket is attached to the bottom side of the magnet to facilitate sealing. A second magnet is placed on the backside of the chip to provide interfacial force to prevent leakage [129].

For higher pressure resistance tests, a threaded mating port for a commercially available capillary fitting can be directly fabricated into an injection-molded polymer chip [43]. Burst pressures can achieve approximately 10 MPa using this elegant solution. However, the relatively large footprint limits the maximum port density. Steel hypodermic needles using both frictional interference fits and threaded fittings are often adopted for fluid interconnects in thermoplastic microfluidic chips and have been successfully demonstrated for high-pressure applications. The resulting interfaces offer pressure resistance on the order of 40 MPa with low dead volumes within the flow path.

2.4 Fluidic Control Components

2.4.1 Valve

Fluidic pumps and valves can be integrated into IVD devices to provide more accurate and functional flow and reaction manipulation on a small scale. In particular, valves play an important role in controlling the behavior of small amount of fluids for solution storage and preservation, fluid guidance, and sequential reagent delivery for chemical and biomedical applications. They are necessary for many IVD devices that integrate multiple functions such as sample preparation, complex assays that include incubation, mixing, or reaction steps, and quantitative result outputs [143].

Microvalves have been categorized into passive [144–146] and active [17, 116, 139, 147, 148, 149, 150, 151, 152, 153] types for IVD applications. Passive valves usually take advantage of innovative geometric design and modification of surface energy [144–146]. Their operation is simple, and they can be operated at low actuation pressure. For example, the valving mechanism of lab-on-a-disk used temporary capillary stop valves to stop flow at low spinning speeds and open them to fluid passage with increasing rotational velocity [154]. Active valves are useful for shutoff or proportional valving using magnetic [151], piezoelectric [148], electrostatic [147], phase-change [116, 150, 153], mechanical [139], shape-memory [149], or pneumatic [17, 152] actuation mechanisms. Among these, pneumatic PDMS valves [91] fabricated using soft lithographic methods [47] represent one of the most used classes of valve technology for cell biology-based IVD applications (Fig. 2.10a) [155–157]. The mechanism of pneumatic PDMS valve modulation of fluidic flow rate is the applied gas pressure used to deform a thin elastomeric membrane to occlude an underlying, vertically arranged microchannel. In order to simplify the control of PDMS valve systems, several groups have developed elastomeric valves actuated by alternate mechanisms including braille display and machine screw [144, 145, 158]. The moving pins of braille display can act as linear actuators for programmable valves and pumps integrated with multiple layers of PDMS membranes (Fig. 2.10b) [158]. In addition, small machine screws may be cast into a PDMS substrate directly above a membrane-enclosed channel, allowing the elastomer membrane to be compressed and close the underlying channel by simply rotating said screws (Fig. 2.10c) [144, 154]. A similar concept is an indirect screw-assisted hydraulic valve in which the screw actuator can be used to provide the pressure applied to a water-filled PDMS control channel that permits the opening and adjustment of the underlying microchannel width [145]. Based on deformable elastomer microfluidic systems (NanoFlex valve), Fluidigm is the current largest commercial LOC technology company on the market for life science applications such as singe-cell genomics and chip-based, high-throughput PCR.

While PDMS valves are at high density and capable of performing as a pump because of their deformable nature, they are fundamentally limited because the

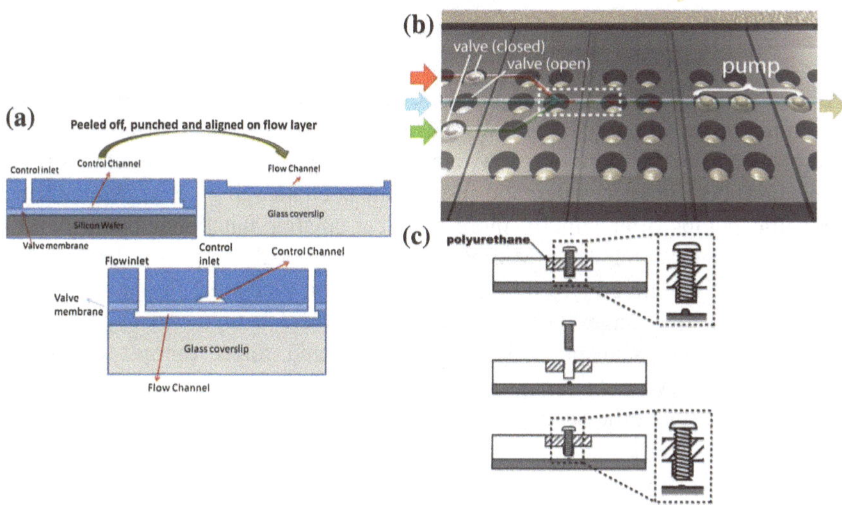

Fig. 2.10 **a** Intermediate fabrication steps (*top*) and cross section (*bottom*) of the three-layer PDMS valve integration. There are three types of actuation mechanisms utilized to enclose the PDMS channels, they are gas, **b** braille display, and **c** screws, respectively [154, 158, 159]

material properties can impact the mechanical integrity, surface chemistry, pressure limits, gas permeability, and solvent compatibility of the resulting IVD system. On the other hand, thermoplastics such as PC, PMMA, and COP offer compatibility with high-throughput replication methods and multiple surface modification options and offer dimensional stability and rigidity suitable for high-pressure applications. However, thermoplastics cannot support the integration of deformable valves due to their rigidity. A proportional PDMS valve combined with PDMS and COP was proposed for high-pressure applications [159]. This hybrid valve employs a small PDMS cylinder cast at the base of a threaded access port, with a small-gauge threaded needle used to controllably rotate and deform the PDMS cylinder into an underlying microchannel. This hybrid microfluidic valve retains the mechanical integrity and surface chemistry of thermoplastic substrates, and it reduces the gas permeability and solvent compatibility by limiting the area of PDMS exposed to the microchannels.

As an alternative to PDMS valves, thermally responsive gel valves have been successfully integrated with thermoplastic chips [60, 116, 153, 160]. These porous polymer monoliths are act as valves that operate through temperature-controlled swelling and deswelling of the temperature-responsive poly(N-isopropylacrylamide) (P[NIPAAm]) matrix. The leakage pressure for these valves is approximately 10 MPa, and the temperature-control mechanism can be operated by an optic switch for response time in the range of several seconds [116]. This is a particularly good example of an on-chip, high-pressure binary valve.

2.4.2 Pump

External pumps such as syringe pumps, diaphragm pumps, and peristaltic pumps are common in research-based IVD devices [113, 114]. They offer the advantage of precise and continuous flow rate control; nevertheless, the integration of a pump into the compact systematic instrument is challenging [161]. Miniaturized peristaltic pumps offer slightly more practical smaller functional parts for IVD pumping systems [162, 163]. Microfabricated reciprocal and rotary displacement pumps can be integrated into fluidic cartridges to provide large flow rate but wide operation pressure ranges [164]. In addition to elegant high-end systems, low-cost pumping mechanisms include human-powered finger pumps, chemically induced pressure/vacuum pumps, and spring-based pumps that can be built into IVD devices without the need for any electrical power supply [161]. Furthermore, silicone tubing and a plastic device such as a micropinch valve can be used for RT-PCR-based HIV detection devices [98]. Electrokinetical methods including electroosmotic pumping, capillary electrophoresis, and electrochromatography have been used for fluidic movement and charged species separation. These methods have also been adopted for diagnostic immunoassays [165, 166]. Traditional DC-powered systems that require high voltages for fluidic movement may damage the sample, but low-powered AC-based systems such as electrothermal and electroosmosis pumps can minimize this problem and are more suitable for IVD applications [167, 168].

Several commercial diagnostic products based on such mechanisms have already been developed. Micronics introduced access cards to separate and capture an analyte from a complex biological specimen for further analysis [169]. Fluidigm, previously mentioned in our section on valving, uses PDMS-based fluidic networks controlled by arrays of pneumatic valves to provide both valving and peristaltic pumping to precisely control on-chip liquid flow for single-cell biology and sample identification applications (Fig. 2.11a) [170].

Spinning disklike IVD devices employ centrifugal, capillary, and Coriolis forces to sequentially transport samples and reagents [171–174]. The Piccolo

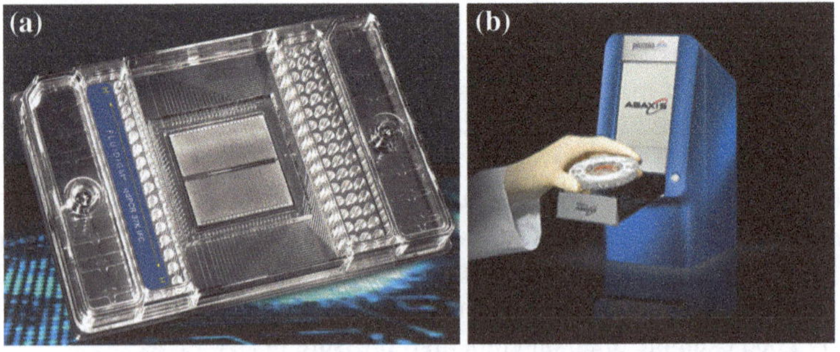

Fig. 2.11 **a** Image of integrated fluidic circuit (IFC) chip. **b** The Piccolo Xpress POC clinical system (https://www.fluidigm.com, http://www.piccoloxpress.com)

Xpress is a portable clinical diagnostic system for on-site patient testing of lipid and liver panels from whole blood, serum, or plasma (Fig. 2.11b) [175]. In spinning disk-type devices, fluids can be pumped toward the rim of the disk at a different flow rates by manipulating spin speed, surface chemical properties, and various chamber and channel geometric details. Separating plasma from whole blood, fluidic mixing, metering of liquid, and signal enhancement can also be integrated into such systems, which have been applied for rapid detection of glucose, hemoglobin, and alcohol in human whole blood [176]. Other sensing platforms can be combined with the fluidic CD format, such as carbon electrodes for dielectrophoresis [177] and whole blood immunoassay [174, 178].

The electrowetting-on-dielectric (EWOD) method is an approach in which droplet movement is controlled by electrical forces. Multiple functions such as generation, mixing, sorting, and splitting of droplets can be controlled by a network of electrodes covered with a dielectric coating. The ability to precisely control the movement of small amounts of liquid offers many potential applications in IVD diagnostics [179, 180]. Rapid immunoassay, on-chip sample extraction, and rapid PCR from whole blood samples with a handheld instrument and disposable chips can be achieved [181]. In another EWOD-based approach, Advanced Liquid Logic uses silicone oil to encase whole blood sample droplets to avoid the protein surface adsorption problem for whole blood sample analysis.

2.4.3 Mixer

Rapid and efficient mixing is critical for sample dilution, dried regent resuspension, and multiple reagent reactions in IVD devices [182]. Mixing in microfluidic platforms is difficult because Reynolds numbers are low (<1), flow is laminar, and mixing is dominated by diffusion only. The Reynolds number is explained by the following formula:

$$\mathrm{Re} = \frac{\rho U L}{\mu}$$

In this formula, ρ is the density of fluid, U is the mean velocity of the object relative to the fluid, L is the characteristic linear dimension, and μ is the dynamic viscosity of the fluid.

In order to resolve this issue, efficient micromixing has been achieved via a number of active and passive mixing mechanisms [169, 183]. The difference between active and passive mixing is based on the need for external forces or not to achieve efficient mixing. In active mixing, external driving forces such as acoustic wave agitation, magnetic manipulation of magnetic beads, or air bubbles have been applied to enhance sample mixing. Several mixer designs have been successfully incorporated into IVD devices. Membrane-based micromixers use air to expand and compress a series of chambers; they create creating mixing in fluidic channels as part of a platform for DNA extraction, leukocyte purification, and genotyping from whole blood [184]. Air is also used to actuate multilayer PDMS chambers for mixing in urine analysis devices [185]. Typical acoustic mixing uses an external piezoelectric transducer to oscillate bubbles trapped along the side walls of microchannels to produce rapid mixing [186]. In addition, a lead zirconate titanate (PZT) disk can be coupled with laser-cut PSA-based mixing chamber to achieve rapid mixing using trapped air bubbles [187]. Centrifugal forces are not only used for pumping, separation, and detection, but they can also be used for on-chip mixing. In droplet-based microfluidic platforms, mixing is achieved by simply merging individual droplets into one [180, 181, 188].

In passive mixing, liquids are driven through delicately designed microstructures, such as a modified Tesla structure, which divides a flow into two streams that collide from opposite directions to increase contact area between the different streams [189], or a staggered herringbone mixer (SHM) to induce chaotic mixing (Fig. 2.12) [190, 191]. A patterned staggered herringbone topography on surfaces can be used to generate chaotic flows in contexts other than pressure-driven flows in microchannels. Chaotic flows exist in the laminar shear flow at the boundary layer of an extended flow over the staggered herringbone structure surface. As a result, this stirring of the boundary layer enhances the rates of diffusion-limited reactions at surfaces.

Splitting and recombining liquid streams with microfabricated 3D structures can be adopted to enhance mixing in microchannels [192, 193]. In addition, a combination of baffles and curved channels can be used to enhance mixing [194, 195]. Improvement in mixing efficiency can also be achieved through modification

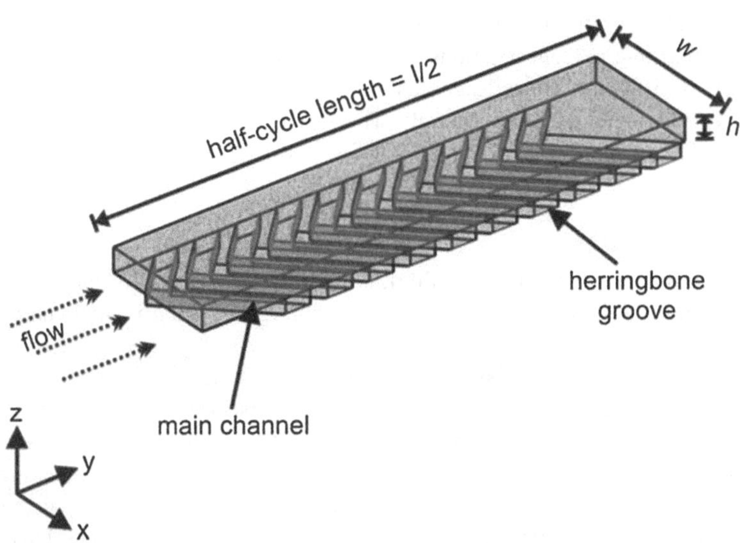

Fig. 2.12 **a** Schematic diagram of a-half cycles of the SHM. The direction of asymmetry of the each half-cycle of herringbones switches with respect to thecenterline of the channel from one region to the next [191]

of channel surface energy such as through the use of a checkerboard pattern of blended hydrophobic and hydrophilic regions in microchannels [196]. Complicated porous geometry, for example, photopolymerized polymer monoliths in microchannels, is another approach to enhance mixing and improve overall chemical reaction efficiency [197]. The applications of mixers are easily coupled with other actuation and sensing mechanisms for pathogen DNA detection and various clinical diagnostics [76, 198].

In addition to polymeric materials, mixing is also demonstrated on a paper-based IVD platform by stacking paper strips in a flat Y-geometry mixer [199], and cotton threads with knots have been used for routing and mixing for body fluidic analyses [200].

2.5 Applications

Body fluids continuously circulate throughout the human body to deliver necessary nutrients and transport metabolic waste. They can reveal abundant information about the health status of an individual [201]. The molecular constituents in body fluids are directly associated with the physiological state of the body. Therefore, detection of target molecules in body fluids is useful for revealing information for prevention, identification, and treatment of a variety of diseases. The most common traditional technologies for molecular diagnostics include ELISA, PCR, and mass spectrometry platforms (MS) [202]. However, these technologies are limited to laboratory use because they rely on complicated sample purification, sophisticated instruments, intensive time, and labor and require highly trained personnel. For these reasons, the challenge remains to develop affordable, specific, sensitive, user-friendly, and rapid technologies for IVD molecular diagnostics. The following section will discuss polymer-based IVD applications in the molecular diagnostic field.

2.5.1 Sample Preparation

Proper sample preparation can enhance signal intensity and reduced background signal as a means of improving sensitivity. Preparation encompasses sample filtration, purification, concentration, diffusion, and fractionation of target analytes from background matrices. Although large numbers of IVD devices can process whole blood samples to obtain diagnostic results directly, assays can be more accurate and sensitive with adequate sample preparation.

Cell lysis is a valuable approach for obtaining cell-relative disease information [203, 204]. Lysis can be achieved on-chip by a variety of approaches including mechanical and chemical methodologies. Functional beads, when incorporated with other actuation techniques, are effective and worthwhile tools for cell lysis

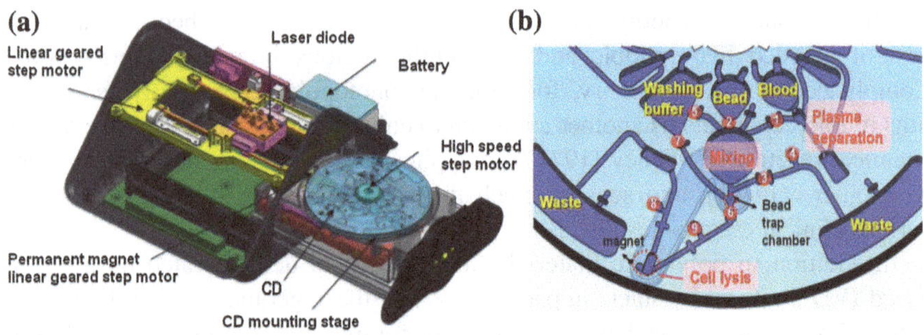

Fig. 2.13 **a** Schematic diagram of the portable lab-on-a-disc device. **b** The detailed microfluidic layout and functions of the polymeric disk [206]

studies. A disposable bead-based blender can be used for thick-walled bacillus spores and mycobacterium cell lysis for molecular diagnostics [205]. Similarly, a laser can be used to irradiate magnetic particles for rapid lysis aimed at the extraction of DNA from *Escherichia coli* and the hepatitis B virus (Fig. 2.13) [206].

Nucleic acid purification is important for DNA-based diagnostics that require signal amplification without an increase in news. In conventional PCR diagnostics, sample preparation steps such as DNA extraction are time-consuming procedures that require a commercially available kit and experienced technical skill. Recent developments have overcome these barriers by incorporating integrated sample purification steps for on-chip DNA extraction [184, 207, 208]. Integrated LOC platforms use magnetic beads to purify and concentrate leukocytes prior to DNA extraction in lysis solution; this is followed by PCR amplification [184, 207].

Capture and purification of RNA using silica beads has been accomplished for *E. coli* RNA and viral RNA from influenza-A (H1N1)-infected mammalian cells [209] with real-time nucleic acid sequence-based amplification (NASBA) for target sequence amplification [210]. A combination of chemical and mechanical hybrid methods for bacterial cells lysis has also been reported [211]. In this process, bacterial samples pass through a porous polymer monolith containing detergent lytic matrix, resulting in a concentrated DNA that is eluted for PCR diagnostics. Thermochemical lysis using lysis buffer at high temperatures has been adopted for on-chip protein-based detection of *Bacillus subtilis* cells and spores [212]. Electroporation, a molecular biology technique in which an electrical field is applied to cells in order to increase cell membrane permeability, has been adopted for on-chip disruption of cell membranes and cell lysis [212–214]. Cell lysis has also been performed using an optically induced electrical field to induce transmembrane potential for fibroblasts and oral cancer cell lysis [215].

Microdroplet technology offers another choice for analyzing intracellular proteins using electrical lysis followed by incubation with antibody-conjugated beads for further analysis [216]. In addition to cell lysis, a droplet-extraction system can be used to extract acidic analytes from urine; this is followed by capillary electrophoresis and laser-induced fluorescence detection [217]. The passive isolation of T lymphoma cells can be achieved using PEG droplets to encapsulate dextran droplets within a microchannel and partitioning the cells into the PEG phase as they remain in the aqueous droplet [218].

2.5.2 Separation

The core of lab-on-a-chip devices is closely linked to separations of biochemical species on chip. Separation is important for IVD devices because it removes interfering agents and isolates them from the target analytes prior to detection. This enhances the dynamic detection range to properly target and seek out molecular information [162, 219].

Separation techniques such as liquid-phase extraction, solid-phase extraction (SPE), liquid chromatography, capillary electrophoresis (CE), dielectrophoresis (DEP), isoelectric focusing (IEF), micellar electrokinetic chromatography (MEKC), isotachophoresis (ITP), optical tweezers, magnetic manipulation and capture, acoustic waves and fields, size-based filtration using nanostructures and microstructures filters, and various combinations of flow behaviors have previously been adopted [20, 219-229]. The parameters that can be leveraged for separation include surface charge, polarizability, mass, size, pH value, and various physical, chemical, and immune-conjugation interactions between materials and molecules that can be combined to achieve the best separation performance. The integrative character of LOC systems allows for the performance of on-chip sample preparation and serial separation for IVD applications. For example, polymer microfluidic chips employing in situ photopolymerized polymethacrylate monoliths for high-performance liquid chromatography separations of peptides have been described [30]. In this procedure, the integrated chip design employs a 15-cm-long separation column containing a reversed-phase polymethacrylate monolith as a stationary phase, with its front end seamlessly coupled to a 5-mm-long methacrylate monolith that functions as a solid-phase extraction (SPE) element for sample cleanup and enrichment, serving to increase both detection sensitivity and separation performance (Fig. 2.14a).

For whole blood analysis, accurate, fast, and affordable IVD systems are important in clinical diagnostics. It is important to separate red blood cells from plasma due to the interference caused by the cellular components of blood. Microfluidic devices for continuous, real-time blood plasma separation are introduced to achieve the operation with simple and portable manners.

A chip-based blood analyzer that allows for the separation of plasma from whole blood samples based on gravitational sedimentation of red and white cells into filtering trenches has been reported [229]. Another example is adopting Zweifach–Fung effect to separate blood plasma and blood cells from whole blood (Fig. 2.14b) [231]. In corporation with the on-chip plasma separation, an integrated blood barcode chip that can sensitively sample a large panel of protein biomarkers over broad concentration ranges and within 10 min of sample

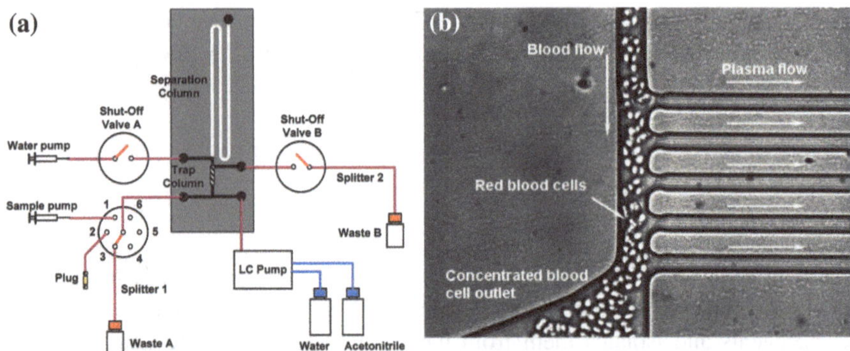

Fig. 2.14 **a** Experimental system for online sample cleanup/enrichment–HPLC separations, with an integrated SPE trap column used for online sample cleanup and enrichment. **b** Scheme depicting plasma separation from a finger prick of blood by harnessing the Zweifach–Fung effect. Multiple DNA-encoded antibody barcode arrays are patterned within the plasma-skimming channels for in situ protein measurements [30, 232]

collection time has been achieved [232]. This process enables on-chip blood separation according to the Zweifach–Fung effect and rapid measurement of a panel of plasma proteins from quantities of whole blood as small as those obtained by a finger prick. Cells can be separated by flowing blood through a low-flow-resistance primary channel with high-resistance, centimeter-long channels that branch off at right angles. As the resistance ratio is increased between the branches and the primary channel, a critical streamline moves closer to the primary channel wall adjoining the branch channels. Blood cells with a radius larger than the distance between this critical streamline and the primary channel wall are directed away from the high-resistance channels, and ~15 % of the plasma is skimmed into the high-resistance channels. The remaining whole blood is directed toward a waste outlet.

Centrifugal microfluidics provides on-chip red blood cell separation from plasma according to differences in cell density and centrifugally driven sedimentation [171, 178, 233]. Diffusion phenomena have been used in combination with flow in microchannels to separate smaller molecules from other larger ones in whole blood that diffuses more slowly [234]. Another application of such technology is the immunoassay-based detection of IgG in whole blood samples (rat) using a PDMS-cellulose composite film [235]. Based on the principle of cross-flow, the authors demonstrated that their microfluidic device could isolate plasma from raw samples.

Antibodies are usually used to select target cells in miniaturized devices. They can be bound to gold nanoparticle (AuNP) surfaces or magnetic nanoparticle surfaces for highly efficient cell purification and selection [236, 237]. An electrical field can be applied for high binding efficiencies between cells and magnetic beads in the droplet [238]. Bacteria such as salmonella and *Staphylococcus aureus* can also be captured and separated using magnetic beads coated with antibodies [227].

In addition, DEP can be used for continuous enrichment of platelets from diluted whole blood based on the size difference of platelets to other blood cells [239].

2.5.3 Reagent Storage

Stable storing of reagents and biomolecules for biological events using IVD devices is critical for complicated on-site blood, saliva, or urine assays [185]. Storage can be either wet or dry format according to the stability and storage duration of selected biomolecule species. There are only a few examples of stored wet-type reagents. The majority are lab-on-a-disk platforms that integrate CD-laser-openable valves [240], or pre-storage of liquid reagents in glass ampoules for DNA extraction on a fully integrated lab-on-a-chip cartridge [241]. No liquid loss is observed for ethanol and H_2O stored for 300 days at room temperature. Applicability of this concept is demonstrated by performing DNA extraction after 140 days of reagent pre-storage. DNA yield from 32 mL of whole blood was up to 199 ng, which is 77 % of an off-chip reference extraction.

On-chip storage of dry reagents is well developed for IVD tests. Immobilization of proteins on beads improves dry reagent storage capacity [114, 242]. LFA strips adopt dried AuNP-conjugated antibody regents at the conjugation pad for rapid pregnancy, drug abuse, and other diagnostic tests [220, 243]. Flow-through immunoassays are another option that can be performed using on-chip reagent storage that employs a porous membrane or porous monoliths patterned with capture molecules to capture biomolecules for further tests [111, 115].

The other most popular case of reagent storage is in the use of glucose sensors that include dried glucose oxidase and electron transfer catalysts to create an electrical signal after glucose-induced reduction–oxidation reactions. Antibodies are often stored in a dried or lyophilized state that does not cause unrecoverable loss of activity before recondition for further tests in the future [161, 244]. In order to extend the stability and retention of action, sugar is frequently used to spike the biomolecules. Dextran, for example, covalently coupled to amino groups, thus providing a coating with a low contact angle suitable for antibody immobilization. The contact angle measurement shows that dextran-coated chips are stable for more than two months, which enables production of large batches that can be stored for extended periods of time. By using the dextran-coated pillar-based platform, a C-reactive protein (CRP) assay was successfully implemented by immobilizing αCRP antibody in 1 % trehalose solution [245]. In addition, the gold–antibody-conjugated, sugar-dried matrices can retain 80–96 % of their activity after 60 days of storage at elevated temperatures [114].

A different approach for solution storage is blister pack technology, which employs the vacuum deposition of a thin film of metal on the storage chamber to prevent significant permeability to water vapor during storage [246].

2.5.4 Detection of Metabolites and Small Molecules

Metabolites are the intermediates and products of metabolism that provide various functions including fuel, structure, signaling, stimulatory, and inhibitory effects on enzymes, and interactions with other organisms by processing nutrients [247]. Metabolites are major diagnostic indicators of disease, state-of-health monitoring, and drug abuse tests. Among the thousands of target molecules in the human body, glucose, cholesterol, triglycerides, creatinine, lactate, ammonia, and urea are the most often targeted metabolites when using IVD diagnostics [248]. One of the best-known clinical IVD analyzers is the i-STAT handheld system for in vitro quantitative diagnostic measurement of blood gases (pH, $pC0_2$, $p0_2$) and whole blood electrolytes (sodium, Na^+; potassium, K^+, and ionized calcium, iCa^{2+}) in the neonatal and pediatric intensive care units from whole blood or plasma (Fig. 2.15) [249]. To perform a test, two to three drops of blood obtained using the finger-prick method are applied to a cartridge, which is then inserted into the i-STAT handheld device. Prior to running a test, each cartridge initiates a series of preset quality control diagnostics to monitor the quality of the sample and validate the reagent. Each test cartridge contains chemically sensitive biosensors on a silicon chip that are configured to perform diagnostic indicator tests related to disease state and clinical practice guidelines.

The best-known metabolite-related diagnostic measurement is that of blood glucose, which can present diagnosis and management values for diabetes mellitus, an endocrine disorder afflicting more than 125 million people worldwide. The biosensors for glucose sensing account for approximately 85 % of the entire biosensor market [250]. In the 1990s, diabetic complications were found to be controllable with tight regulation of glucose levels by patients, prompting the

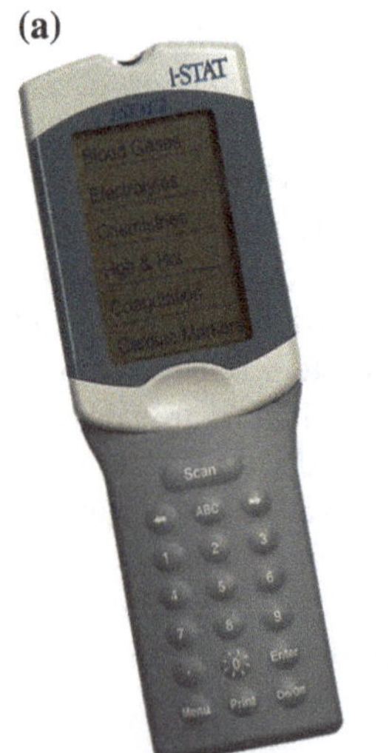

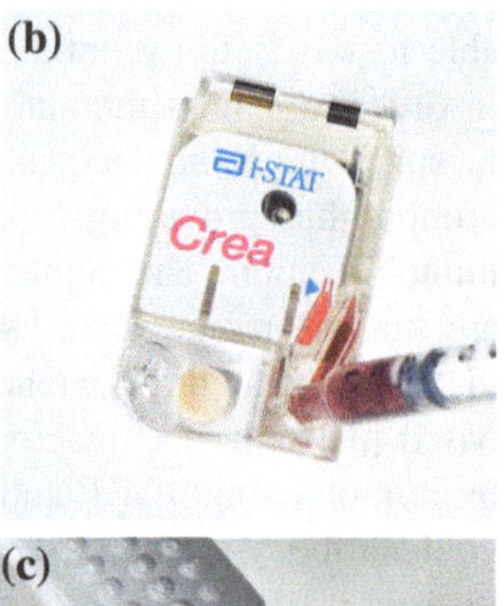

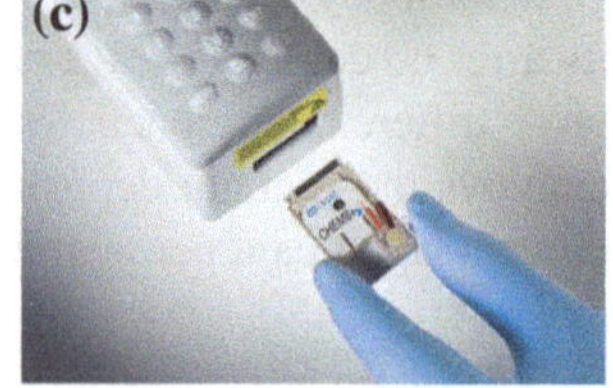

Fig. 2.15 **a** Image of i-STAT handheld detector and **b**, **c** different functional cartridges for in vitro quantitative measurement of electrolytes, hematology, blood gas, coagulation, and cardiac marker from whole blood or plasma (www.abbottpointofcare.com)

development of IVD glucose sensors [251]. The first glucose sensor for home use was announced in the early 1980s. Current glucose meters rely primarily on measuring the electrochemical signals of blood absorbed by test strips via capillary force [249].

According to new regulations that require more accurate sensing results, researchers and scientists are trying to invent more accurate and rapid sensing platforms with novel materials and designs. For example, colorimetric glucose sensors based on intrinsic peroxidase-like activity of graphene, carbon nanotubes, and other metal nanomaterials have recently been developed [252–255]. The results of these ventures appear promising, but more clinical trials and FDA approval are required before practical usage.

In addition to glucose, cholesterol, triglycerides, and other plasma lipids, screening is crucial in the management of cardiovascular disease [256, 257]. Stroke and diabetes are correlated to high levels of cholesterol. Creatinine, a by-product of kidney function and a breakdown product of creatine phosphate in muscle, is produced at a constant rate in healthy individuals. Measuring serum creatinine is a simple test and is the most commonly used indicator of renal function through estimation of the glomerular filtration rate [258].

Renal dysfunction, liver function, and asthma status can often be detected through the measurement of ammonia levels [259]. High ammonia levels in blood are related to slow conversion to urea due to liver impairment [260]. High levels of breath ammonia are also indicators of uremia and kidney failure [261] and could act as a target molecule for noninvasive IVD breath sensing diagnostics [262, 263].

2.5.5 DNA- and RNA-Based Diagnosis

Nucleic acids which include DNA and RNA are polymeric macromolecules or large biological molecules essential for carrying genetic information [264–267]. They are made from monomers known as nucleotides that are composed of a 5-carbon sugar, a phosphate group, and a nitrogenous base.

PCR and numerous other methods of selectively amplifying pre-synthesized and purified nucleotide sequences are always adopted for DNA- or RNA-based assays based on the fact that they are exceptionally sensitive and highly selective. However, laboratory-based PCR is expensive and time-consuming and requires bulky equipment; IVD-format DNA assays do not. In IVD-format assays, the test sample volumes are in microliters, so sample volume cost can be dramatically reduced, and heating/cooling cycles can be performed more rapidly. Examples of POC nucleic acid test platforms that are commercially available or close to market are listed in Table 2.2 [265].

In PCR, sensitivity is controlled by hybridization efficiency. High hybridization efficiency results in low background signal caused by mismatched or spurious DNA sequences. The binding kinetics and target specificity of PCR can be modulated by various factors, such as ionic strength, hybridization number, reaction

Table 2.2 Example POC nucleic acid test platforms that are commercially available or close to market [265]

Platform	Manufacturer	Simple prep included?	Amplification	Detection	Time to result (min)[a]	Web site
GeneXpert	Cepheid	Y	PCR	RTF	<120	www.cepheid.com
Liat Analyzer	IQuum	Y	PCR	RTF	<60	www.iquum.com
MDx	Biocartis	Y	PCR	RTF	Unknown	www.biocartis.com
FL/ML	Enigma	Y	PCR	RTF	<45	www.enigmadiagnostics.com
FilmArray	Idaho technologies	Y	PCR	RTF	60	www.idahotech.com
Razor	Idaho technologies	N	PCR	RTF	<60	www.idahotech.com
R.A.P.I.D.	Idaho technologies	N	PCR	RTF	<30	www.idahotech.com
LA-200	Eileen	N	Isothermal (LAMP)	RTT	<60	www.eiken.co.jp
Twista	TwistDX	N	Isothermal (RPA)	RTF	<20	www.twistdx.co.uk
BART	Lumora	N	Isothermal (LAMP)	RTB	<60	www.lumora.co.uk/
Genie II	Optigene	N	Isothermal (LAMP)	RTF	<20	www.optigene.co.uk
SAMBA	Diagnostics for the real world	N	Isothermal (similar to NASBA)	NALF	>60	Not available
BESt Cassette[b]	BioHelix/UstarBiotech	N	Not included, but typically isothermal	NALF	N/A	www.biohelix.com; www.bioustar.com

[a]Time to result depends upon the particular assay. Longer times may be required for assays with a reverse transcriptase step
[b]Device sold by BioHellx ill the USA; manufactured and sold by Ustar Biotech in China
Abbreviations: RTB, real-time bioluminescence; *RTF,* real-time fluorescence; and *RTT,* real-time turbidimetry

temperature, and probe density [269, 270]. Lower probe densities can lead to higher hybridization efficiencies and more rapid binding kinetics; however, the smaller measured signal may introduce background interference.

Circulating nucleic acid molecules are released from dying cells when they break down [271]. Analysis of circulating DNA fragment size and composition shows promise for cancer and prenatal diagnostics [271, 272]. A single-molecule spectroscopy technique combined with microfluidic system has been developed to analyze DNA biomarkers directly in serum [273]. This system allows quantification of circulating DNA in volumes of serum <1 pL without additional DNA isolation or an enzymatic amplification step.

Identification of DNA mutations that modulate the effectiveness of biological reagents on target-specific pathways is highly valuable for cancer treatment [274]. Tumor-specific mutations can be found in the circulating DNA extracted from the serum of cancer patients. Tagged-amplicon deep sequencing (TAm-Seq) has been developed to amplify and sequence large genomic regions from as little as a single copy of circulating DNA [274, 275]. It has been found that the 5995 screened genomic bases for low-frequency mutations identify mutations present in circulating DNA as low as 2 % with sensitivity and specificity >97%. Methylated DNA is another promising biomarker in urine or blood that can be used for early detection of cancer and other genetic diseases [276, 277]. Quantitation of cellular DNA methylation levels in colon cancer patients has shown to be a promising alternative to colonoscopy for detecting colorectal cancer [278].

Current infectious disease diagnostic methods employ mainly sputum smear microscopy, immunoassay, and culture of bacilli and molecular species diagnostics [279–284]. A biopsy specimen positive for smear microscopy is a cost-effective way to obtain results in minutes. However, this identification method has low sensitivity and specificity. Bacillus cultures are considered the standard method for infectious disease detection in a clinical center. Therefore, results cannot be obtained instantly based on the fact that bacillus requires weeks of culture for proper diagnosis. Immunoassay can be accomplished in minutes depending on incubation times and format [181]. However, direct detection of disease status by IVD immunoassay often suffers from inadequate limits of detection. Due to the majority of newly infected cases in developing countries, a rapid, simple, low cost, and highly accurate on-site detection platform using PCR-based molecular diagnostics for infectious disease diagnosis is vitally important for early detection.

Identification of multiple subtyping of influenza virus using an integrated PDMS device has been achieved [285]. This device was able to extract viral RNA from clinical samples using specific nucleotide probes conjugated on the surface of magnetic beads. The extracted RNA was amplified by one-step RT-PCR, and the products were optically detected with a TaqMan® fluorescence system. Experimental results demonstrated that detection of influenza virus subtypes could be completed as quickly as 110 min. Expedient diagnosis of HIV-1 using a microfluidic array has been demonstrated [286]. Researchers in this endeavor equipped their PDMS microfluidic chip array with a sample treatment system and a nucleic acid amplification stage and integrated the device with functional analytical steps including cell lysis, DNA extraction, PCR, and optical detection. Their device was able to detect DNA fragments from an HIV-infected Jurkat T cell line within 95 min.

A PC chip device that could rapidly detect influenza A/H1N1 virus in human clinical specimens has been developed [287]. In this approach, PC chips were fabricated using a multilayer injection molding technique in which three plastic layers were used to construct the inlet, outlet, fluidic channel, and reaction chambers with dimensions of approximately 72 × 25 × 1.5 mm. A reaction volume of on 15 mL was required. The efficiency of these plastic devices was confirmed using novel primer sets specifically targeted to the hemagglutinin (HA) gene of influenza A/H1N1 and clinical specimens. Eighty-five human clinical swab samples were tested using real-time PCR, the results of which demonstrated 100 % sensitivity and specificity (72 positive and 13 negative cases). Another PC platform was developed that used a disposable and valveless flow cell that supported PCR coupled with microarray hybridization in the same chamber [288]. This device worked by confining liquid to a reaction chamber during thermocycling until additional wash buffer moved the reagent on to the waste chamber. This approach allowed the authors to produce 300 copies of bacterial DNA without the need for disassembly or specialized instrumentation.

Xpert MTB/RIF (Cepheid, Sunnyvale, CA, USA) is one of the most powerful commercial products for TB control in resource-limited settings. It is an automated molecular test for Mycobacterium tuberculosis (MTB) and resistance to rifampin (RIF), and it uses heminested real-time PCR assay to amplify an MTB-specific sequence of the rpoB gene, which is probed with molecular beacons for mutations within the rifampin-resistance determining region [288]. The MTB/RIF test provided sensitive detection of tuberculosis and rifampin resistance directly from untreated sputum in less than two hours with minimal hands-on time.

MicroRNAs (miRNA) are small noncoding RNA molecules (19–22 nucleotides in length) that are frequently dysregulated in cancer and are promising biomarkers for cancer classification and prognosis monitoring [289]. A PDMS-based microRNA detection with laminar flow-assisted dendritic amplification has been realized [290]. This approach permitted specific sequences at a limit of detection of 0.5 pM from a 0.5 μL sample solution with a detection time of 20 min. In addition, a nanopore sensor was developed for detection of miRNAs in samples of plasma from lung cancer patients. This sensor can quantitate cancer-associated miRNAs at subpicomolar concentrations, as well as distinguish single mismatch sequence [291]. However, the commercially available IVD test for miRNA is still being investigated.

2.5.6 Protein-Based Diagnosis

Proteins are large biological molecules consisting of one or more long chains of amino acid residues. They are necessary for various biological functions, and processes include catalyzing metabolic reactions, replicating DNA, responding to stimuli, transporting molecules from one location to another, and tissue repair [292]. Based on a proteomics study, levels of specific protein biomarkers directly reflect disease stages and have been regarded as one of the most promising target molecules for disease diagnosis. The traditional diagnostic in clinical center is using MS or ELISA [293, 294]. Nevertheless, MS is too expensive, and laboratory-based ELISA involves too many manual processes for use in IVD tests.

Current protein analyses are mainly based on the ELISA format. The best-known home IVD protein detection device is a LFA-format test for pregnancy test. The detection mechanism in these tests is capture of the pregnancy hormone human chorionic gonadotropin (hCG) on thin film. Many IVD devices use immunoassay technology based on antigen–antibody-binding events to target disease-specific protein markers such as prostate-specific antigen (PSA) for cancer, carcinoembryonic antigen (CEA) for colorectal cancer, human epidermal growth factor receptor 2 (HER2) for breast cancer, glycated hemoglobin (HbA1c) for diabetics, C-reactive protein (CRP) for inflammation including cardiovascular disease, and troponin I for cardiac damage [295–301]. Colorimetric or fluorescent readout signals are then used to quantitatively visualize protein-binding events to specific recognition molecules [12]. A multiplexed detection platform comprising a PDMS microfluidic microparticle array containing gel-based microstructures has been proposed [302]. This platform can detect two protein tumor markers, hCG and PSA. In this microfluidic arrangement, spherical biofunctionalized polystyrene microbeads were incorporated onto polyacrylamide gel microstructures. Detection of hCG and PSA was based on a binding assay in serum samples with LOD below the cutoff values for the diagnosis of cancer. Using low-density polyethylene, Chou et al. [303] employed a hot-embossing mold replicated from an anisotropically etched silicon wafer to an aluminum epoxy crating to fabricate efficient through-hole microarrays. In this research, the authors demonstrated that they could monitor the essential inflammatory biomarker, CRP, thus fortifying the idea that such materials could be used for real-world clinical measurements.

Despite the development of numerous ELISA-based technologies for on-chip protein sensing, sensitivity, multiplicity, quantification, portability, turnaround time, readout signal, and cost remain challenge for IVD platforms. A microfluidic chip (mChip) for on-chip immunoassay for HIV diagnosis and multiplex protein assays has been proposed to solve the abovementioned issues. This mChip integrates fluid handling and silver reduction to obtain highly sensitive immunoassay results within 30 min [127, 304]. Furthermore, a volumetric bar chart chip (V-Chip) has been introduced for visual quantification of biomarkers based on volumetric measurements of oxygen generated to push color ink upward on an mChip [305]. On-chip immunoassay starts with sliding the upper plate over the lower, and the advance of ink in each individual channel indicates the amount of produced oxygen, which correlates to the concentration of the corresponding ELISA target protein. The visible bar chart is used to obtain quantitative results for data processing without additional instrumentation. Although the V-Chip uses a glass slide as the substrate, a plastic substrate may be used to reduce device cost. In addition,

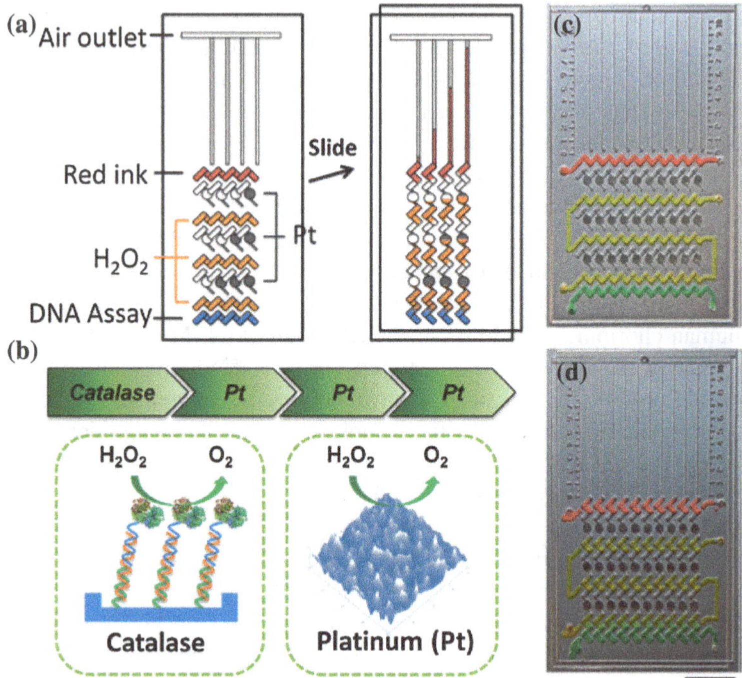

Fig. 2.16 Working principle of the multistage propelled Vchip (MV-Chip). **a** Schematic illustration of a MV-Chip for DNA assays. Platinum amplification (black circles) show larger bar chart advancement than those without amplification in the detection units. **b** Propelling mechanism of the MV-Chip. Catalase introduced by DNA hybridization is the initiator and the platinum films amplify the signals. **c, d** The representative flow path of each reagent before and after an oblique slide. The red lane, yellow lane, and green lane represent ink, H2O2 and DNA assay, respectively. Scale bar is 1 cm for (c) and (d) [253]

sensitivity could be further amplified by pre-depositing multistage uniform platinum films in the device for performing cascade amplification and 20 pM DNA targets after three stages of platinum-catalyzed propulsion can be detected. As a result, single-nucleotide polymorphism and multiplex DNA detection were carried out to demonstrate this powerful application (Fig. 2.16) [306].

Improvement of sensitivity is critical for low-concentration disease biomarker detection. Sensitivity can be enhanced by introducing new signal amplification approaches into ELISA or possessing more binding sites for enzymes conjugate. The plasmonic effect of AuNPs has been adopted and incorporated into an ELISA for colorimetric detection of proteins [307]. In addition, electrostatic interactions between green fluorescent proteins (GFP) and AuNPs have been used for protein detection in undiluted serum [293]. Furthermore, proteins can be detected via colorimetric or light scattering signals based on the aggregation of antibody-conjugated AuNPs in the presence of target protein [308, 309]. The colorimetric detection results caused by surface plasmon resonance can be observed by the naked eye without additional equipment.

2.5.7 Cell Analysis

Identification and enumeration of specific cells such as microbes, viruses, and parasites are critical for disease diagnosis [294, 295, 310]. The assay of cells in IVD format is based on specific antibodies, proteins, or aptamers used to capture corresponding cells. These antibodies, proteins, or aptamers have enzymes, nanomaterials, or signal molecules conjugated to them to produce detectable or visible signal for quantitative or quasi-quantitative detection [311]. Bacterial and virus assays can use antibodies to capture whole or fragmented organisms for diagnosis.

A 3D PDMS microfluidic chip with an integrated array of microwells capable of long-term tumor spheroid cultivation and the evaluation of anticancer drug activity has been developed [312]. By culturing spheroids of HT-29 human carcinoma cells onto a microfluidic chip over the course of four weeks, introducing a cytostatic drug (5-fluorouracil), and incubating on the PDMS chip, the cytotoxic effect on HT-29 cells could be evaluated. The authors observed cell death based on decreasing spheroid diameter. An intriguing PDMS platform showed that a device could exhibit the capacity for self-loading in order to determine the minimum inhibitory concentration (MIC) of antibiotics against bacteria [313]. In this research, bacterial growth in microfluidic chambers in the presence of a pH indicator produced a visible colorimetric change detectable under ambient light. Based on this principle, the authors measured the MIC of vancomycin, tetracycline, and kanamycin against *Enterococcus faecalis* 1131, *Proteus mirabilis* HI4320, *Klebsiella pneumoniae*, and *E. coli* MG1655.

Circulating tumor cells (CTCs), cells that shed from the primary tumor site and present in the bloodstream, provide a rare model system for potentially understanding the mechanism of cancer development. CTCs are rare in the bloodstream, so it is a challenge to capture and count them if the sample amount is limited. A microfluidic platform enabling long-term proliferation of captured CTCs from single cells into colonies has been proposed [314]. The researchers in this endeavor opted for a hydrogel culture system to avoid cell encapsulation, which bears the risk of cell loss in a 3D environment that allows for spatial confinement and protection of single cells. Prostate cancer cells can be captured on a transparent PDMS CTC-chip and subsequently encapsulated into a synthetic biomimetic hydrogel matrix facilitating their clonal expansion to spheroids (Fig. 2.17) [314].

Incorporating an immunomagnetic assay and a fluidic system, CellSearch, a FDA-approved screening system, has been developed to enumerate CTCs in samples of whole blood. In this process, a 7.5 mL sample of blood is centrifuged to

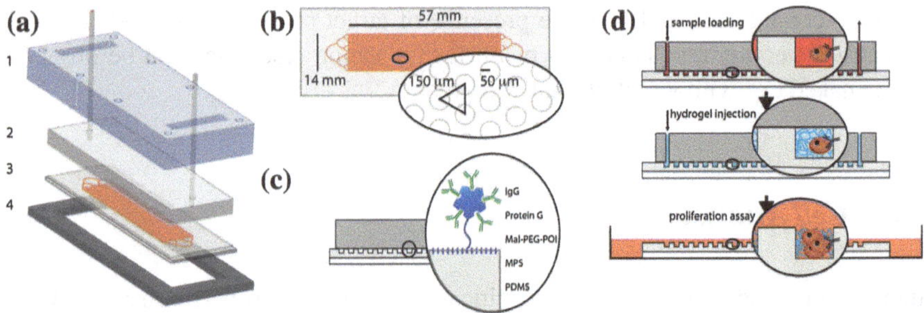

Fig. 2.17 Versatile CTC capture and culture platform. **a** Microfluidic setup. (*1*) Transparent PMMA cover, (*2*) PDMS cover with inlet and outlet tubing, (*3*) microstructured PDMS on a glass slide, and (*4*) metal support frame. **b** Micropillar geometry. **c** PDMS functionalization principle. PEG-linkers connected to Protein G are covalently bound to MPS-treated PDMS; IgGs of interest are added and fixed to Protein G with paraformaldehyde. **d** Schematic of an experiment. In the first step, the sample is flushed through the chip and CTCs captured; then, the chip is washed and hydrogel injected, and finally, the gel-encapsulated cells are transferred to cell culture conditions [314]

separate solid blood components from plasma and then, ferrofluid nanoparticles with antibodies that target epithelial cell adhesion are added to magnetically separate CTCs from the bulk of other cells in the blood. After staining, using monoclonal antibodies and DNA sequences, CTCs are scanned to enumerate cell number. In addition to CTCs, magnetic particles/beads coated with specific antibodies for antigens expressed on cancer cells are used to tag target cells for magnetic capture [315-317].

References

1. Blumenthal D (2009) N Engl J Med 360:1477–1479
2. Myers FB, Lee LP (2008) Lab Chip 8:2015–2031
3. Woolf SH (2008) JAMA 299:211–213
4. Beaudet AL, Belmont JW (2008) Annu Rev Med 59:113–129
5. Zhang Y, Ozdemir P (2009) Anal Chim Acta 638:115–125
6. Herold KE, Rasooly A (2009) Lab on a chip technology: fabrication and microfluidics. Horizon Scientific Press
7. Cui HF, Ye JS, Chen Y, Chong SC, Sheu FS (2006) Anal Chem 78:6347–6355
8. Li N, Tourovskaia A, Folch A (2003) Crit Rev™ Biomed Eng 31
9. Huh D, Matthews BD, Mammoto A, Montoya-Zavala M, Hsin HY, Ingber DE (2010) Science 328:1662–1668
10. Whitesides GM (2006) Nature 442:368–373
11. Holland CA, Kiechle FL (2005) Curr Opin Microbiol 8:504–509
12. Gervais L, De Rooij N, Delamarche E (2011) Adv Mater 23:H151–H176
13. Chin CD, Linder V, Sia SK (2012) Lab Chip 12:2118–2134
14. Tüdős AJ, Besselink GA, Schasfoort RB (2001) Lab Chip 1:83–95
15. Sharma H, Nguyen D, Chen A, Lew V, Khine M (2011) Ann Biomed Eng 39:1313–1327

16. Tsao CW, DeVoe DL (2009) Microfluid Nanofluid 6:1–16
17. Unger MA, Chou HP, Thorsen T, Scherer A, Quake SR (2000) Science 288:113–116
18. Clarson SJ, Semlyen JA, Clarson SJ (1993) Siloxane polymers. Prentice Hall, Englewood Cliffs
19. Van Krevelen DW, Te Nijenhuis K (2009) Properties of polymers: their correlation with chemical structure; their numerical estimation and prediction from additive group contributions. Elsevier, Amsterdam
20. Lötters J, Olthuis W, Veltink P, Bergveld P (1997) J Micromech Microeng 7:145
21. Mata A, Fleischman AJ, Roy S (2005) Biomed Microdevices 7:281–293
22. Anderson JR, Chiu DT, Wu H, Schueller OJ, Whitesides GM (2000) Electrophoresis 21:27–40
23. Hillborg H, Ankner J, Gedde UW, Smith G, Yasuda H, Wikström K (2000) Polymer 41:6851–6863
24. Duffy DC, McDonald JC, Schueller OJA, Whitesides GM (1998) Anal Chem 70:4974–4984
25. Lee JN, Park C, Whitesides GM (2003) Anal Chem 75:6544–6554
26. Nunes PS, Ohlsson PD, Ordeig O, Kutter JP (2010) Microfluid Nanofluid 9:145–161
27. Niles WD, Coassin PJ (2008) Assay Drug Dev Technol 6:577–590
28. Liedert R, Amundsen LK, Hokkanen A, Mäki M, Aittakorpi A, Pakanen M, Scherer JR, Mathies RA, Kurkinen M, Uusitalo S (2012) Lab Chip 12:333–339
29. Bilenberg B, Hansen M, Johansen D, Özkapici V, Jeppesen C, Szabo P, Obieta I, Arroyo O, Tegenfeldt J, Kristensen A (2005) J Vac Sci Technol, B 23:2944–2949
30. Liu J, Chen CF, Tsao CW, Chang CC, Chu CC, DeVoe DL (2009) Anal Chem 81:2545–2554
31. Lv Y, Lin Z, Svec F (2012) Anal Chem 84:8457–8460
32. Huang WJ, Chang FC, Chu PPJ (2000) Polymer 41:6095–6101
33. Khanarian G, Celanese H (2001) Opt Eng 40:1024–1029
34. Keyes DL, Lamonte RR, McNally D, Bitritto M (2001) Photonics Spectra 35
35. Saaem I, Ma KS, Marchi AN, LaBean TH, Tian J (2010) ACS Appl Mater Interf 2:491–497
36. Shi YW, Wang Y, Abe Y, Matsuura Y, Miyagi M, Sato S, Taniwaki M, Uyama H (1998) Appl Opt 37:7758–7762
37. Yamazaki M (2004) J Mol Catal A: Chem 213:81–87
38. Shiono T (2011) Polym J 43:331–351
39. Shin JY, Park JY, Liu C, He J, Kim SC (2005) Pure Appl Chem 77:801–814
40. Schulz U (2006) Appl Opt 45:1608–1618
41. Jena R, Chester S, Srivastava V, Yue C, Anand L, Lam Y (2011) Sens Actuators B: Chem 155:93–105
42. Esch MB, Kapur S, Irizarry G, Genova V (2003) Lab Chip 3:121–127
43. Mair DA, Geiger E, Pisano AP, Fréchet JM, Svec F (2006) Lab Chip 6:1346–1354
44. Moreau WM (1988) Semiconductor lithography. Springer, Berlin
45. Stryer L, Amy TLT, Solas D (1991) Light-directed, spatially addressable parallel
46. Weibel DB, DiLuzio WR, Whitesides GM (2007) Nat Rev Microbiol 5:209–218
47. Xia Y, Whitesides GM (1998) Annu Rev Mater Sci 28:153–184
48. Xia Y, McClelland JJ, Gupta R, Qin D, Zhao XM, Sohn LL, Celotta RJ, Whitesides GM (1997) Adv Mater 9:147–149
49. Zhao XM, Xia Y, Whitesides GM (1996) Adv Mater 8:837–840
50. King E, Xia Y, Zhao XM, Whitesides GM (1997) Adv Mater 9:651–654
51. Rogers JA, Paul KE, Jackman RJ, Whitesides GM (1997) Appl Phys Lett 70:2658–2660
52. Childs WR, Nuzzo RG (2002) J Am Chem Soc 124:13583–13596
53. Jeon S, Menard E, Park JU, Maria J, Meitl M, Zaumseil J, Rogers JA (2004) Adv Mater 16:1369–1373
54. Kumar A, Whitesides GM (1993) Appl Phys Lett 63:2002–2004
55. Xu Q, Rioux RM, Dickey MD, Whitesides GM (2008) Acc Chem Res 41:1566–1577
56. Chen CS, Mrksich M, Huang S, Whitesides GM, Ingber DE (1997) Science 276:1425–1428

57. Backofen U, Matysik FM, Lunte CE (2002) Anal Chem 74:4054–4059
58. Gates BD, Xu Q, Stewart M, Ryan D, Willson CG, Whitesides GM (2005) Chem Rev 105:1171–1196
59. Chung K, Crane MM, Lu H (2008) Nat Methods 5:637–643
60. Zorlutuna P, Annabi N, Camci-Unal G, Nikkhah M, Cha JM, Nichol JW, Manbachi A, Bae H, Chen S, Khademhosseini A (2012) Adv Mater 24:1782–1804
61. Love JC, Ronan JL, Grotenbreg GM, van der Veen AG, Ploegh HL (2006) Nat Biotechnol 24:703–707
62. Taylor AM, Blurton-Jones M, Rhee SW, Cribbs DH, Cotman CW, Jeon NL (2005) Nat Methods 2:599–605
63. Zheng B, Tice JD, Ismagilov RF (2004) Adv Mater 16:1365–1368
64. Xia Y, Rogers JA, Paul KE, Whitesides GM (1999) Chem Rev 99:1823–1848
65. Von Philipsborn AC, Lang S, Bernard A, Loeschinger J, David C, Lehnert D, Bastmeyer M, Bonhoeffer F (2006) Nat Protoc 1:1322–1328
66. Love JC, Estroff LA, Kriebel JK, Nuzzo RG, Whitesides GM (2005) Chem Rev 105:1103–1170
67. George J (1992) Preparation of thin films. CRC Press, Boca Raton
68. Lee KH, Su YD, Chen SJ, Tseng FG, Lee GB (2007) Biosens Bioelectron 23:466–472
69. Anderson BB, Brodsky AM, Burgess LW (1997) Langmuir 13:4273–4279
70. Stewart ME, Mack NH, Malyarchuk V, Soares JA, Lee TW, Gray SK, Nuzzo RG, Rogers JA (2006) Proc Natl Acad Sci 103:17143–17148
71. German RM (1990) Powder injection molding. Cambridge University Press, Cambridge
72. Malloy RA (1994) Plastic part design for injection molding. Hanser Gardner Publications
73. Heckele M, Schomburg W (2004) J Micromech Microeng 14:R1
74. Mela P, van den Berg A, Fintschenko Y, Cummings EB, Simmons BA, Kirby BJ (2005) Electrophoresis 26:1792–1799
75. Piotter V, Hanemann T, Ruprecht R, Hausselt J (1997) Microsyst Technol 3:129–133
76. Choi SH, Kim DS, Kwon TH (2009) Microsyst Technol 15:309–316
77. Vasconcelos PV, Lino FJ, Baptista AM, Neto RJ (2006) Wear 260:30–39
78. Ito H, Kazama K, Kikutani T (2007) Effects of process conditions on surface replication and higher-order structure formation in micromolding
79. Angelov A, Coulter J (2008) Polym Eng Sci 48:2169–2177
80. Østergaard PF, Matteucci M, Reisner W, Taboryski R (2013) Analyst 138:1249–1255
81. Gadegaard N, Mosler S, Larsen NB (2003) Macromol Mater Eng 288:76–83
82. Becker H, Heim U (2000) Sens Actuators, A 83:130–135
83. Fiorini GS, Jeffries GD, Lim DS, Kuyper CL, Chiu DT (2003) Lab Chip 3:158–163
84. Cameron NS, Roberge H, Veres T, Jakeway SC, Crabtree HJ (2006) Lab Chip 6:936–941
85. Kricka LJ, Fortina P, Panaro NJ, Wilding P, Alonso-Amigo G, Becker H (2002) Lab Chip 2:1–4
86. Koerner T, Brown L, Xie R, Oleschuk RD (2005) Sens Actuators B: Chem 107:632–639
87. Greener J, Li W, Ren J, Voicu D, Pakharenko V, Tang T, Kumacheva E (2010) Lab Chip 10:522–524
88. Wu JT, Chang WY, Yang SY (2010) J Micromech Microeng 20:075023
89. Fu XX, Kang XN, Zhang B, Xiong C, Jiang XZ, Xu DS, Du WM, Zhang GY (2011) J Mater Chem 21:9576–9581
90. Gerlach G, Dotzel W (2008) Introduction to microsystem technology: a guide for students. Wiley, New York
91. Chou SY, Krauss PR, Zhang W, Guo L, Zhuang L (1997) J Vac Sci Technol, B 15:2897–2904
92. Guo LJ (2007) Adv Mater 19:495–513
93. Nilsson D, Balslev S, Kristensen A (2005) J Micromech Microeng 15:296
94. Gustafsson O, Mogensen KB, Kutter JP (2008) Electrophoresis 29:3145–3152
95. Srinivasan R, Braren B (1989) Chem Rev 89:1303–1316

96. Srinivasan R, Braren B, Casey KG, Yeh M (1989) Appl Phys Lett 55:2790–2791
97. Johnson TJ, Waddell EA, Kramer GW, Locascio LE (2001) Appl Surf Sci 181:149–159
98. Lee SH, Kim SW, Kang JY, Ahn CH (2008) Lab Chip 8:2121–2127
99. Sauer-Budge AF, Mirer P, Chatterjee A, Klapperich CM, Chargin D, Sharon A (2009) Lab Chip 9:2803–2810
100. Chuag SH, Chen GH, Chou HH, Shen SW, Chen CF (2013) Sci Technol Adv Mater 14:044403
101. Dhokia V, Kumar S, Vichare P, Newman S, Allen R (2008) Proceedings of the Institution of Mechanical Engineers. Part B: J Eng Manuf 222:137–157
102. Jung W, Han J, Kai J, Lim JY, Sul D, Ahn CH (2013) Lab Chip 13:4653–4662
103. Coltro WKT, de Jesus DP, da Silva JAF, do Lago CL, Carrilho E (2010) Electrophoresis 31:2487–2498
104. do Lago CL, da Silva HDT, Neves CA, Brito-Neto JGA, da Silva JAF (2003) Anal Chem 75:3853–3858
105. Duarte GRM, Price CW, Augustine BH, Carrilho E, Landers JP (2011) Anal Chem 83:5182–5189
106. Duarte GRM, Coltro WKT, Borba JC, Price CW, Landers JP, Carrilho E (2012) Analyst 137:2692–2698
107. de Souza FR, Alves GL, Coltro WKT (2012) Anal Chem 84:9002–9007
108. Oliveira KA, de Oliveira CR, da Silveira LA, Coltro WKT (2013) Analyst 138:1114–1121
109. Kim AR, Kim JY, Choi K, Chung DS (2013) Talanta 109:20–25
110. Wu H, Huang B, Zare RN (2005) Lab Chip 5:1393–1398
111. Ramachandran S, Singhal M, McKenzie KG, Osborn JL, Arjyal A, Dongol S, Baker SG, Basnyat B, Farrar J, Dolecek C (2013) Diagnostics 3:244–260
112. Heller A, Feldman B (2010) Acc Chem Res 43:963–973
113. Yager P, Edwards T, Fu E, Helton K, Nelson K, Tam MR, Weigl BH (2006) Nature 442:412–418
114. Stevens DY, Petri CR, Osborn JL, Spicar-Mihalic P, McKenzie KG, Yager P (2008) Lab Chip 8:2038–2045
115. Lafleur L, Stevens D, McKenzie K, Ramachandran S, Spicar-Mihalic P, Singhal M, Arjyal A, Osborn J, Kauffman P, Yager P (2012) Lab Chip 12:1119–1127
116. Chen G, Svec F, Knapp DR (2008) Lab Chip 8:1198–1204
117. Abgrall P, Low LN, Nguyen NT (2007) Lab Chip 7:520–522
118. Kameoka J, Craighead HG, Zhang H, Henion J (2001) Anal Chem 73:1935–1941
119. Bhattacharyya A, Klapperich CM (2007) Lab Chip 7:876–882
120. Liston E, Martinu L, Wertheimer M (1993) J Adhes Sci Technol 7:1091–1127
121. Chen C, Liu J, Hromada L, Tsao C, Chang C, DeVoe D (2009) Lab Chip 9:50–55
122. Ogilvie I, Sieben V, Floquet C, Zmijan R, Mowlem M, Morgan H (2010) J Micromech Microeng 20:065016
123. Su YC, Lin L (2005) IEEE Trans Adv Packag 28:635–642
124. Yussuf A, Sbarski I, Hayes J, Solomon M, Tran N (2005) J Micromech Microeng 15:1692
125. Lei KF, Ahsan S, Budraa N, Li WJ, Mai JD (2004) Sens Actuators, A 114:340–346
126. Hitzbleck M, Delamarche E (2013) Chem Soc Rev 42:8494–8516
127. Chin CD, Laksanasopin T, Cheung YK, Steinmiller D, Linder V, Parsa H, Wang J, Moore H, Rouse R, Umviligihozo G (2011) Nat Med 17:1015–1019
128. Huh D, Kim HJ, Fraser JP, Shea DE, Khan M, Bahinski A, Hamilton GA, Ingber DE (2013) Nat Protoc 8:2135–2157
129. Atencia J, Cooksey GA, Jahn A, Zook JM, Vreeland WN, Locascio LE (2010) Lab Chip 10:246–249
130. Zhong JF, Chen Y, Marcus JS, Scherer A, Quake SR, Taylor CR, Weiner LP (2008) Lab Chip 8:68–74
131. Bhagat AAS, Jothimuthu P, Pais A, Papautsky I (2007) J Micromech Microeng 17:42
132. Fredrickson CK, Fan ZH (2004) Lab Chip 4:526–533

133. Bings NH, Wang C, Skinner CD, Colyer CL, Thibault P, Harrison DJ (1999) Anal Chem 71:3292–3296
134. Gray B, Jaeggi D, Mourlas N, Van Drieenhuizen B, Williams K, Maluf N, Kovacs G (1999) Sens Actuators, A 77:57–65
135. Pan T, Baldi A, Ziaie B (2006) J Microelectromech Syst 15:267–272
136. Murphy ER, Inoue T, Sahoo HR, Zaborenko N, Jensen KF (2007) Lab Chip 7:1309–1314
137. Renzi RF, Stamps J, Horn BA, Ferko S, VanderNoot VA, West JA, Crocker R, Wiedenman B, Yee D, Fruetel JA (2005) Anal Chem 77:435–441
138. Brivio M, Oosterbroek RE, Verboom W, van den Berg A, Reinhoudt DN (2005) Lab Chip 5:1111–1122
139. Yin H, Killeen K, Brennen R, Sobek D, Werlich M, van de Goor T (2005) Anal Chem 77:527–533
140. Ro KW, Liu J, Knapp DR (2006) J Chromatogr A 1111:40–47
141. Saarela V, Franssila S, Tuomikoski S, Marttila S, Östman P, Sikanen T, Kotiaho T, Kostiainen R (2006) Sens Actuators B: Chem 114:552–557
142. Mohanty S, Beebe D, Mensing G Chips Tips
143. Linder V (2007) Analyst 132:1186–1192
144. Hulme SE, Shevkoplyas SS, Whitesides GM (2009) Lab Chip 9:79–86
145. Zheng Y, Dai W, Wu H (2009) Lab Chip 9:469–472
146. Lee KS, Ram RJ (2009) Lab Chip 9:1618–1624
147. Teymoori MM, Abbaspour-Sani E (2005) Sens Actuators, A 117:222–229
148. Yang EH, Lee C, Mueller J, George T (2004) J Microelectromech Syst 13:799–807
149. Kohl M, Dittmann D, Quandt E, Winzek B (2000) Sens Actuators, A 83:214–219
150. Beebe DJ, Moore JS, Bauer JM, Yu Q, Liu RH, Devadoss C, Jo BH (2000) Nature 404:588–590
151. Chen H, Gu W, Cellar N, Kennedy R, Takayama S, Meiners JC (2008) Anal Chem 80:6110–6113
152. Grover WH, von Muhlen MG, Manalis SR (2008) Lab Chip 8:913–918
153. Hisamoto H, Funano SI, Terabe S (2005) Anal Chem 77:2266–2271
154. Weibel DB, Kruithof M, Potenta S, Sia SK, Lee A, Whitesides GM (2005) Anal Chem 77:4726–4733
155. Thorsen T, Maerkl SJ, Quake SR (2002) Science 298:580–584
156. Fan HC, Wang J, Potanina A, Quake SR (2011) Nat Biotechnol 29:51–57
157. EmreáAraci I (2012) Lab Chip 12:2803–2806
158. Gu W, Zhu X, Futai N, Cho BS, Takayama S (2004) Proc Natl Acad Sci USA 101:15861–15866
159. Chen CF, Liu J, Chang CC, DeVoe DL (2009) Lab Chip 9:3511–3516
160. Peters EC, Svec F, Fréchet JM (1997) Adv Mater 9:630–633
161. Weigl B, Domingo G, LaBarre P, Gerlach J (2008) Lab Chip 8:1999–2014
162. Yobas L, Tang KC, Yong SE, Ong EKZ (2008) Lab Chip 8:660–662
163. Xie J, Shih J, Lin Q, Yang B, Tai YC (2004) Lab Chip 4:495–501
164. Laser D, Santiago J (2004) J Micromech Microeng 14:R35
165. Escarpa A, González MC, López Gil MA, Crevillén AG, Hervás M, García M (2008) Electrophoresis 29:4852–4861
166. Gao Y, Sherman PM, Sun Y, Li D (2008) Anal Chim Acta 606:98–107
167. Sin ML, Gao J, Liao JC, Wong PK (2011) J Biol Eng 5:1–22
168. Huang SB, Wu MH, Lee GB (2009) Sens Actuators B: Chem 142:389–399
169. Kokoris M, Nabavi M, Lancaster C, Clemmens J, Maloney P, Capadanno J, Gerdes J, Battrell C (2005) Methods 37:114–119
170. Melin J, Quake SR (2007) Annu Rev Biophys Biomol Struct 36:213–231
171. Gorkin R, Park J, Siegrist J, Amasia M, Lee BS, Park JM, Kim J, Kim H, Madou M, Cho YK (2010) Lab Chip 10:1758–1773
172. Mark D, Metz T, Haeberle S, Lutz S, Ducrée J, Zengerle R, von Stetten F (2009) Lab Chip 9:3599–3603

173. Kim J, Kido H, Rangel RH, Madou MJ (2008) Sens Actuators B: Chem 128:613–621
174. Lee BS, Lee YU, Kim HS, Kim TH, Park J, Lee JG, Kim J, Kim H, Lee WG, Cho YK (2011) Lab Chip 11:70–78
175. Park H, Ko DH, Kim JQ, Song SH (2009) Korean J Lab Med 29:430–438
176. Steigert J, Grumann M, Brenner T, Mittenbühler K, Nann T, Rühe J, Moser I, Haeberle S, Riegger L, Riegler J (2005) J Assoc Lab Autom 10:331–341
177. Martinez-Duarte R, Gorkin RA III, Abi-Samra K, Madou MJ (2010) Lab Chip 10:1030–1043
178. Lee BS, Lee JN, Park JM, Lee JG, Kim S, Cho YK, Ko C (2009) Lab Chip 9:1548–1555
179. Teh SY, Lin R, Hung LH, Lee AP (2008) Lab Chip 8:198–220
180. Malic L, Brassard D, Veres T, Tabrizian M (2010) Lab Chip 10:418–431
181. Sista R, Hua Z, Thwar P, Sudarsan A, Srinivasan V, Eckhardt A, Pollack M, Pamula V (2008) Lab Chip 8:2091–2104
182. Du Y, Zhang Z, Yim C, Lin M, Cao X (2010) Biomicrofluidics 4:024105
183. Lee CY, Chang CL, Wang YN, Fu LM (2011) Int J Mol Sci 12:3263–3287
184. Lien KY, Liu CJ, Lin YC, Kuo PL, Lee GB (2009) Microfluid Nanofluid 6:539–555
185. Lin CC, Wang JH, Wu HW, Lee GB (2010) J Assoc Lab Autom 15:253–274
186. Ahmed D, Mao X, Juluri BK, Huang TJ (2009) Microfluid Nanofluid 7:727–731
187. Nath P, Fung D, Kunde YA, Zeytun A, Branch B, Goddard G (2010) Lab Chip 10:2286–2291
188. Srinivasan V, Pamula VK, Fair RB (2004) Lab Chip 4:310–315
189. Hong CC, Choi JW, Ahn CH (2004) Lab Chip 4:109–113
190. Stroock AD, Dertinger SK, Ajdari A, Mezić I, Stone HA, Whitesides GM (2002) Science 295:647–651
191. Williams MS, Longmuir KJ, Yager P (2008) Lab Chip 8:1121–1129
192. Kim DS, Lee SH, Kwon TH, Ahn CH (2005) Lab Chip 5:739–747
193. Tofteberg T, Skolimowski M, Andreassen E, Geschke O (2010) Microfluid Nanofluid 8:209–215
194. Tsai RT, Wu CY (2011) Biomicrofluidics 5:014103
195. Long M, Sprague MA, Grimes AA, Rich BD, Khine M (2009) Appl Phys Lett 94:133501
196. Swickrath MJ, Burns SD, Wnek GE (2009) Sens Actuators B: Chem 140:656–662
197. Mair DA, Schwei TR, Dinio TS, Svec F, Fréchet JM (2009) Lab Chip 9:877–883
198. Jung JH, Kim GY, Seo TS (2011) Lab Chip 11:3465–3470
199. Osborn JL, Lutz B, Fu E, Kauffman P, Stevens DY, Yager P (2010) Lab Chip 10:2659–2665
200. Reches M, Mirica KA, Dasgupta R, Dickey MD, Butte MJ, Whitesides GM (2010) ACS Appl Mater Interf 2:1722–1728
201. Gutierrez G, Reines HD, Wulf-Gutierrez ME (2004) Critical Care-London 8:373–381
202. Burtis CA, Ashwood ER, Bruns DE (2012) Tietz textbook of clinical chemistry and molecular diagnostics. Elsevier Health Sciences, Amsterdam
203. El-Ali J, Sorger PK, Jensen KF (2006) Nature 442:403–411
204. Nakayama T, Namura M, Tabata KV, Noji H, Yokokawa R (2009) Lab Chip 9:3567–3573
205. Vandeventer PE, Weigel KM, Salazar J, Erwin B, Irvine B, Doebler R, Nadim A, Cangelosi GA, Niemz A (2011) J Clin Microbiol 49:2533–2539
206. Cho YK, Lee JG, Park JM, Lee BS, Lee Y, Ko C (2007) Lab Chip 7:565–573
207. Hoshino K, Huang YY, Lane N, Huebschman M, Uhr JW, Frenkel EP, Zhang X (2011) Lab Chip 11:3449–3457
208. Burtis C, Mailen J, Johnson W, Scott C, Tiffany T, Anderson N (1972) Clin Chem 18:753–761
209. Manage DP, Morrissey YC, Stickel AJ, Lauzon J, Atrazhev A, Acker JP, Pilarski LM (2011) Microfluid Nanofluid 10:697–702
210. Bhattacharyya A, Klapperich CM (2008) Sens Actuators B: Chem 129:693–698
211. Dimov IK, Garcia-Cordero JL, O'Grady J, Poulsen CR, Viguier C, Kent L, Daly P, Lincoln B, Maher M, O'Kennedy R (2008) Lab Chip 8:2071–2078

212. Mahalanabis M, Al-Muayad H, Kulinski MD, Altman D, Klapperich CM (2009) Lab Chip 9:2811–2817
213. Stachowiak JC, Shugard EE, Mosier BP, Renzi RF, Caton PF, Ferko SM, Van de Vreugde JL, Yee DD, Haroldsen BL, VanderNoot VA (2007) Anal Chem 79:5763–5770
214. Lee DW, Cho YH (2007) Sens Actuators B: Chem 124:84–89
215. Lu H, Schmidt MA, Jensen KF (2005) Lab Chip 5:23–29
216. Lin YH, Lee GB (2009) Appl Phys Lett 94:033901
217. Martino C, Zagnoni M, Sandison ME, Chanasakulniyom M, Pitt AR, Cooper JM (2011) Anal Chem 83:5361–5368
218. Sikanen T, Pedersen-Bjergaard S, Jensen H, Kostiainen R, Rasmussen KE, Kotiaho T (2010) Anal Chim Acta 658:133–140
219. Vijayakumar K, Gulati S, Edel JB (2010) Chem Sci 1:447–452
220. Lin CC, Tseng CC, Chuang TK, Lee DS, Lee GB (2011) Analyst 136:2669–2688
221. Dudek MM, Lindahl TL, Killard AJ (2010) Anal Chem 82:2029–2035
222. Pumera M (2007) Electrophoresis 28:2113–2124
223. Freire SL, Wheeler AR (2006) Lab Chip 6:1415–1423
224. Nge PN, Yang W, Pagaduan JV, Woolley AT (2011) Electrophoresis 32:1133–1140
225. Bhagat AAS, Bow H, Hou HW, Tan SJ, Han J, Lim CT (2010) Med Biol Eng Compu 48:999–1014
226. Davies R, Eapen S, Carlisle S (2007) Handb Biosens Biochips
227. Posthuma-Trumpie GA, Korf J, van Amerongen A (2009) Anal Bioanal Chem 393:569–582
228. Qiu J, Zhou Y, Chen H, Lin JM (2009) Talanta 79:787–795
229. Kenyon SM, Meighan MM, Hayes MA (2011) Electrophoresis 32:482–493
230. Hou C, Herr AE (2008) Electrophoresis 29:3306–3319
231. Dimov IK, Basabe-Desmonts L, Garcia-Cordero JL, Ross BM, Ricco AJ, Lee LP (2011) Lab Chip 11:845–850
232. Yang S, Ündar A, Zahn JD (2006) Lab Chip 6:871–880
233. Fan R, Vermesh O, Srivastava A, Yen BK, Qin L, Ahmad H, Kwong GA, Liu CC, Gould J, Hood L (2008) Nat Biotechnol 26:1373–1378
234. Haeberle S, Brenner T, Zengerle R, Ducrée J (2006) Lab Chip 6:776–781
235. Hatch A, Garcia E, Yager P (2004) Proc IEEE 92:126–139
236. Chen X, Zhang LL, Li H, Sun JH, Cai HY, Cui DF (2013) Sens Actuators A-Phys 193:54–58
237. Van de Broek B, Devoogdt N, D'Hollander A, Gijs HL, Jans K, Lagae L, Muyldermans S, Maes G, Borghs G (2011) ACS Nano 5:4319–4328
238. Parolo C, de la Escosura-Muñiz A, Merkoçi A (2013) Biosens Bioelectron 40:412–416
239. Shah GJ, Veale JL, Korin Y, Reed EF, Gritsch HA (2010) Biomicrofluidics 4:044106
240. Pommer MS, Zhang Y, Keerthi N, Chen D, Thomson JA, Meinhart CD, Soh HT (2008) Electrophoresis 29:1213–1218
241. Garcia-Cordero JL, Kurzbuch D, Benito-Lopez F, Diamond D, Lee LP, Ricco AJ (2010) Lab Chip 10:2680–2687
242. Hoffmann J, Mark D, Lutz S, Zengerle R, von Stetten F (2010) Lab Chip 10:1480–1484
243. McKenzie KG, Lafleur LK, Lutz BR, Yager P (2009) Lab Chip 9:3543–3548
244. Li JJ, Ouellette AL, Giovangrandi L, Cooper DE, Ricco AJ, Kovacs GT (2008) IEEE Trans Biomed Eng 55:1560–1571
245. Garcia E, Kirkham JR, Hatch AV, Hawkins KR, Yager P (2004) Lab Chip 4:78–82
246. Jönsson C, Aronsson M, Rundström G, Pettersson C, Mendel-Hartvig I, Bakker J, Martinsson E, Liedberg B, MacCraith B, Öhman O (2008) Lab Chip 8:1191–1197
247. Jokerst JV, Floriano PN, Christodoulides N, Simmons GW, McDevitt JT (2008) Lab Chip 8:2079–2090
248. Clarke JT (2005) A clinical guide to inherited metabolic diseases. Cambridge University Press, Cambridge
249. Luppa PB, Müller C, Schlichtiger A, Schlebusch H (2011) TrAC Trends Anal Chem 30:887–898

250. Lauks IR (1998) Acc Chem Res 31:317–324
251. Wang J (2008) Chem Rev 108:814–825
252. Wilkins E, Atanasov P (1996) Med Eng Phys 18:273–288
253. Song Y, Qu K, Zhao C, Ren J, Qu X (2010) Adv Mater 22:2206–2210
254. Guo Y, Deng L, Li J, Guo S, Wang E, Dong S (2011) ACS Nano 5:1282–1290
255. Song Y, Wang X, Zhao C, Qu K, Ren J, Qu X (2010) Chem-A Eur J 16:3617–3621
256. Su L, Feng J, Zhou X, Ren C, Li H, Chen X (2012) Anal Chem 84:5753–5758
257. Warsinke A (2009) Anal Bioanal Chem 393:1393–1405
258. Rapi S, Bazzini C, Tozzetti C, Sbolci V, Modesti PA (2009) Transl Res 153:71–76
259. Shephard M, Peake M, Corso O, Shephard A, Mazzachi B, Spaeth B, Barbara J, Mathew T (2010) Clin Chem Lab Med 48:1113–1119
260. Wang C, Sahay P (2009) Sensors 9:8230–8262
261. Simon Davies PS, Smith D (1997) Kidney Int 52:223–228
262. Miekisch W, Schubert JK, Noeldge-Schomburg GF (2004) Clin Chim Acta 347:25–39
263. Hibbard T, Killard AJ (2011) Crit Rev Anal Chem 41:21–35
264. Cao W, Duan Y (2007) Crit Rev Anal Chem 37:3–13
265. Lander ES, Linton LM, Birren B, Nusbaum C, Zody MC, Baldwin J, Devon K, Dewar K, Doyle M, FitzHugh W (2001) Nature 409:860–921
266. Niemz A, Ferguson TM, Boyle DS (2011) Trends Biotechnol 29:240–250
267. Lam B, Das J, Holmes RD, Live L, Sage A, Sargent EH, Kelley SO (2013) Nat Commun 4
268. Liong M, Hoang AN, Chung J, Gural N, Ford CB, Min C, Shah RR, Ahmad R, Fernandez-Suarez M, Fortune SM (2013) Nat Commun 4:1752
269. Levicky R, Horgan A (2005) Trends Biotechnol 23:143–149
270. Vainrub A, Pettitt BM (2003) J Am Chem Soc 125:7798–7799
271. Jahr S, Hentze H, Englisch S, Hardt D, Fackelmayer FO, Hesch RD, Knippers R (2001) Cancer Res 61:1659–1665
272. Umetani N, Giuliano AE, Hiramatsu SH, Amersi F, Nakagawa T, Martino S, Hoon DS (2006) J Clin Oncol 24:4270–4276
273. Liu KJ, Brock MV, Shih IM, Wang TH (2010) J Am Chem Soc 132:5793–5798
274. Forshew T, Murtaza M, Parkinson C, Gale D, Tsui DW, Kaper F, Dawson SJ, Piskorz AM, Jimenez-Linan M, Bentley D (2012) Science Transl Med 4 136ra168
275. Murtaza M, Dawson SJ, Tsui DW, Gale D, Forshew T, Piskorz AM, Parkinson C, Chin SF, Kingsbury Z, Wong AS (2013) Nature 497:108–112
276. Warren JD, Xiong W, Bunker AM, Vaughn CP, Furtado LV, Roberts WL, Fang JC, Samowitz WS, Heichman KA (2011) BMC Med 9:133
277. Yu J, Zhu T, Wang Z, Zhang H, Qian Z, Xu H, Gao B, Wang W, Gu L, Meng J (2007) Clin Cancer Res 13:7296–7304
278. Yang Q, Dong Y, Wu W, Zhu C, Chong H, Lu J, Yu D, Liu L, Lv F, Wang S (2012) Nature communications 3:1206
279. Kowalski RP, Karenchak LM, Romanowski EG, Gordon YJ (1999) Ophthalmology 106:1324–1327
280. Ginocchio CC, Zhang F, Manji R, Arora S, Bornfreund M, Falk L, Lotlikar M, Kowerska M, Becker G, Korologos D (2009) J Clin Virol 45:191–195
281. Pabbaraju K, Tokaryk KL, Wong S, Fox JD (2008) J Clin Microbiol 46:3056–3062
282. Stott SL, Hsu CH, Tsukrov DI, Yu M, Miyamoto DT, Waltman BA, Rothenberg SM, Shah AM, Smas ME, Korir GK (2010) Proc Natl Acad Sci 107:18392–18397
283. Hindson BJ, Ness KD, Masquelier DA, Belgrader P, Heredia NJ, Makarewicz AJ, Bright IJ, Lucero MY, Hiddessen AL, Legler TC (2011) Anal Chem 83:8604–8610
284. Lim LS, Hu M, Huang MC, Cheong WC, Gan ATL, Looi XL, Leong SM, Koay ESC, Li MH (2012) Lab Chip 12:4388–4396
285. Wang CH, Lien KY, Hung LY, Lei HY, Lee GB (2012) Microfluid Nanofluid 13:113–123
286. Wang JH, Cheng L, Wang CH, Ling WS, Wang SW, Lee GB (2013) Biosens Bioelectron 41:484–491

287. Song HO, Kim JH, Ryu HS, Lee DH, Kim SJ, Kim DJ, Suh IB, Choi DY, In KH, Kim SW, Park H (2012) PloS ONE 7:e53325
288. Cooney CG, Sipes D, Thakore N, Holmberg R, Belgrader P (2012) Biomed Microdevices 14:45–53
289. Boehme CC, Nabeta P, Hillemann D, Nicol MP, Shenai S, Krapp F, Allen J, Tahirli R, Blakemore R, Rustomjee R (2010) N Engl J Med 363:1005–1015
290. Chen X, Ba Y, Ma L, Cai X, Yin Y, Wang K, Guo J, Zhang Y, Chen J, Guo X (2008) Cell Res 18:997–1006
291. Arata H, Komatsu H, Hosokawa K, Maeda M (2012) PLoS ONE 7:e48329
292. Wang Y, Zheng D, Tan Q, Wang MX, Gu LQ (2011) Nat Nanotechnol 6:668–674
293. Devlin TM (1997) Textbook of biochemistry. Wiley-Liss, New York
294. De M, Rana S, Akpinar H, Miranda OR, Arvizo RR, Bunz UH, Rotello VM (2009) Nat Chem 1:461–465
295. Hu Y, Bouamrani A, Tasciotti E, Li L, Liu X, Ferrari M (2009) ACS Nano 4:439–451
296. Durner J (2010) Angew Chem Int Ed 49:1026–1051
297. Partin A, Yoo J, Carter HB, Pearson J, Chan D, Epstein J, Walsh P (1993) J Urol 150:110–114
298. Benchimol S, Fuks A, Jothy S, Beauchemin N, Shirota K, Stanners CP (1989) Cell 57:327–334
299. Wolff AC, Hammond MEH, Schwartz JN, Hagerty KL, Allred DC, Cote RJ, Dowsett M, Fitzgibbons PL, Hanna WM, Langer A (2006) J Clin Oncol 25:118–145
300. Black S, Kushner I, Samols D (2004) J Biol Chem 279:48487–48490
301. Selvin E, Steffes MW, Zhu H, Matsushita K, Wagenknecht L, Pankow J, Coresh J, Brancati FL (2010) N Engl J Med 362:800–811
302. Adams JE, Bodor GS, Davila-Roman VG, Delmez J, Apple F, Ladenson J, Jaffe A (1993) Circulation 88:101–106
303. Zhu QD, Trau D (2012) Anal Chim Acta 751:146–154
304. Chou J, Du N, Ou T, Floriano PN, Christodoulides N, McDevitt JT (2013) Biosens Bioelectron 42:653–660
305. Gaster RS, Hall DA, Nielsen CH, Osterfeld SJ, Yu H, Mach KE, Wilson RJ, Murmann B, Liao JC, Gambhir SS (2009) Nat Med 15:1327–1332
306. Song Y, Zhang Y, Bernard PE, Reuben JM, Ueno NT, Arlinghaus RB, Zu Y, Qin L (2012) Nat Commun 3:1283
307. Song Y, Wang Y, Qin L (2013) J Am Chem Soc 135:16785–16788
308. de La Rica R, Stevens MM (2012) Nat Nanotechnol 7:821–824
309. Liu X, Dai Q, Austin L, Coutts J, Knowles G, Zou J, Chen H, Huo Q (2008) J Am Chem Soc 130:2780–2782
310. Giljohann DA, Seferos DS, Daniel WL, Massich MD, Patel PC, Mirkin CA (2010) Angew Chem Int Ed 49:3280–3294
311. Clerc O, Greub G (2010) Clin Microbiol Infect 16:1054–1061
312. Basabe-Desmonts L, Ramstrom S, Meade G, O'neill S, Riaz A, Lee L, Ricco A, Kenny D (2010) Langmuir 26:14700–14706
313. Ziolkowska K, Stelmachowska A, Kwapiszewski R, Chudy M, Dybko A, Brzozka Z (2013) Biosens Bioelectron 40:68–74
314. Cira NJ, Ho JY, Dueck ME, Weibel DB (2012) Lab Chip 12:1052–1059
315. Bichsel CA, Gobaa S, Kobel S, Secondini C, Thalmann GN, Cecchini MG, Lutolf MP (2012) Lab Chip 12:2313–2316
316. Miller MC, Doyle GV, Terstappen LW (2009) J Oncol 2010
317. Hoshino K, Chen P, Huang YY, Zhang X (2012) Anal Chem 84:4292–4299

Chapter 3
Low-cost In Vitro Diagnostic Technologies

3.1 Overview of Low-cost In Vitro Diagnostic Technologies

Over the last ten years, increasing interest and research attention have been given to the use of microfluidic devices to carry out diagnostic assays [1, 2]. The evidence shows that such devices are very portable, they require small sample and reagent volumes, and they can conveniently integrate the necessary steps for rapid diagnostic assays. For these reasons, considerable effort and resources have been applied to the development and evaluation of new substrates, microfabrication techniques, and detection methods to develop the least expensive, the most robust, and the most readily disposable and portable in vitro diagnostic devices. In resource-poor and remote settings, such devices could be the key to resolving critical, long-standing human health issues [3]. In vitro diagnostic devices can, in such settings, identify diseases and disease states using an approach that is operable by laypeople with no professional training under conditions of limited power and can do so reliably, accurately, and inexpensively.

Keeping the cost of in vitro diagnostic devices down is particularly important, as it promotes the development, scalability, and deployment of critical in-the-field diagnostic devices, especially for resource-poor settings [4]. It is possible to use approved, inexpensive materials and cost-effective manufacturing to mass produce microfluidic in vitro diagnostic devices. While inexpensive, such devices must still be biocompatible, easily functionalized, adequate for diagnostic detection methods, and non-biohazardous when disposed of.

While the primary substrate materials for microfluidic device fabrication have been silicon and glass, mass production of devices using these materials is cost-prohibitive, especially when targeting a market focused on low-resource communities. Conveniently, plastic and paper both are affordable, versatile, easily disposable, and amenable to mass production of microfluidic devices for diagnostic applications [3, 5–8].

C.-M. Cheng et al., *In-Vitro Diagnostic Devices*, DOI 10.1007/978-3-319-19737-1_3

Here, I will outline the primary advances found in the recently published work (from January 2011 to September 2013). This review will, in particular, cover fabrication methods, designs, capabilities, and practical applications of microfluidic devices for diagnostic applications fabricated with low-cost materials, such as paper, plastic, and thread.

3.2 Paper-Based Microfluidic Devices

As a substrate platform, paper provides distinct advantages for the manufacture of microfluidic devices; it is affordable, abundant, disposable, and amenable to mass production. For these reasons, paper, as a Point-of-care (POC) diagnostic device platform, has become a material of choice [9]. Recent developments in and support for microfluidic paper-based analytical devices (μPADs) for diagnostic applications are summarized in the following sections.

3.2.1 Benefits of Paper

Paper, composed of cellulose, offers many advantages that make it uniquely suitable for POC diagnostics and distinctively compatible for many bioassays [3, 6]: (1) Paper is thin, lightweight, and easily stored; (2) paper is inexpensive (high-quality chromatography paper is $\sim\$6/m^2$) [3]; (3) paper can be easily disposed of by incineration, so it is more eco-friendly than plastic material; (4) paper, being light in color, is suitable for colorimetric assays; (5) paper is thin (0.07–1 mm), so it requires only small sample volumes for wicking; (6) paper is easily modified chemically and allows for biomolecule conjugation (e.g., protein, DNA), so it can be customized to meet very specific diagnostic needs [10]; (7) paper can be patterned with hydrophobic barriers via wax printing (e.g., 96-well and 384-well formats for high-throughput use) [11]; and (8) microchannels can easily be fabricated on paper, and associated microfluidic advantages of paper can be uniquely leveraged [12]. All of these advantageous features indicate that paper is highly suitable for a broad assortment of bioassays including metabolic assays, enzyme-linked immunosorbent assays (paper-based ELISA, or P-ELISA), paper-based cell culture studies, and more [11, 13–26]. Beyond the singular experimental advantages of paper, it is exceptionally advantageous economically, making it distinctly beneficial for use in developing and underdeveloped countries. In this manuscript, we will summarize current studies that use paper bioassay platforms and discuss their benefits to the pharmaceutical industry [1].

3.2.2 Fabrication Techniques

Fabrication of μPADs is typically accomplished by creating reaction wells or zones via specific patterning of hydrophobic barriers and hydrophilic zones onto and into paper. The pattern of wells or zones can, if desired, mimic that of plastic consumables commonly used in laboratory settings, such as 96-well plates. This technique is usually completed by using photoresist and wax [3, 5]. The wax printing method [27], the most frequently used process for quickly producing μPADs, uses a wax ink printer that applies a microfluidic layout directly onto, and ultimately into, a paper surface. The wax-printed paper can then be heated to melt the wax into the paper pores, generating hydrophobic walls within, and not just upon, the paper. Naturally, variations of this approach have been reported by many authors. Using chromatographic paper and an iron template held with a magnet, Songjaroen et al. [28] developed a wax-based fabrication method in which their assembly was dipped into melted wax so that when cooled and the iron template was removed, hydrophilic channels and reaction zones were left patterned on the paper wherever the iron template was in contact. Dungchai et al. [29] developed another process, a screen-printing technique, to deposit wax onto paper in much the same fashion as silk-screen-printing is done for many commercial materials ranging from plastic and glass to paper. Zhang and Zha [30] used a patterned metal template, in this case copper, but they covered the patterned copper template with paraffin film, placed it upon the paper surface, and heated the assembly so that the paraffin diffused into the porous paper to manufacture their patterned hydrophobic walls. Zhong et al. [31] investigated actual substrate material potential, i.e., the potential of different paper types and wax types to produce μPADs. Everything from printing paper, kitchen towels, napkins, and laboratory paper towels were examined as potential microfluidic platforms. For wax material, everything from wax pencils, crayons, candles, and lipsticks were used to generate hydrophobic barriers on and in a variety of paper substrates. Zhong and colleagues examined average pore diameter and permeability as well as fluid-flow mechanics for each paper type. A simple and rapid fabrication process was developed by Nie et al. [32], who elaborated a one-step fabrication process using the ink from a commercial permanent marker and an iron template to pattern microfluidic layouts onto chromatographic paper in less than a minute.

Origami, interestingly enough, was coupled with photolithography and photoresist by Ge et al. [33] and Liu and Crooks [34] to create three-dimensional (3D) μPADs (Fig. 3.1). Patterned paper was folded following a specific sequence to build 3D microfluidic structures. Using this origami technique, 3D μPADs with as many as nine layers could be produced without special tools or tape [34].

Tian et al. [35] showed that the wettability of paper modified with horseradish peroxidase and antibodies (for blood typing) could be recovered by using a plasma treatment process, and the activity of biomolecules could be protected against plasma treatment by using bovine serum albumin (BSA). Lewis et al.

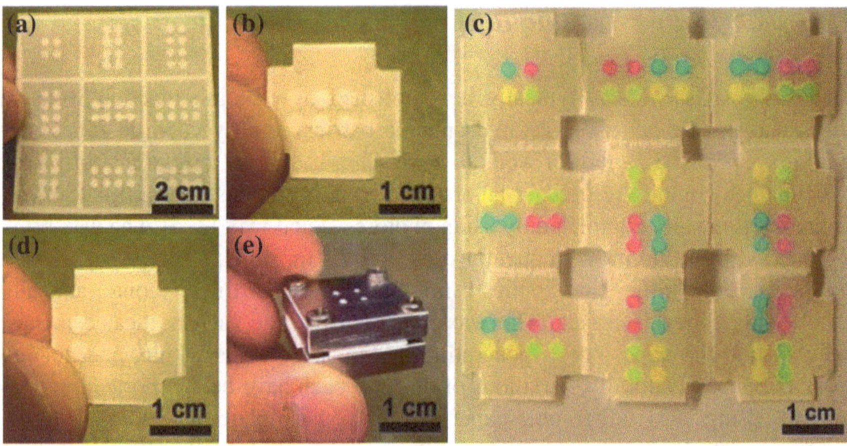

Fig. 3.1 Fabrication process of a 3D μPAD using an origami technique: **a** unfolded patterned chromatographic paper; **b** top and **c** bottom layers of folded patterned paper; **d** 3D μPAD supported by an aluminum clamp; **e** unfolded origami μPAD after wicking of colored solution. Reproduced with permission from Liu and Crooks [34]

[36] developed a rapid, scalable manufacturing for 3D μPADs. Using an adhesive spray, they glued layers of wax-patterned paper together and then cut them to create many individual microdevices. The adhesive spray was shown not to affect colorimetric assays for glucose and protein detection.

He et al. [37] showed that chromatography paper patterning could be achieved by immersing a paper substrate in an octadecyltrichlorosilane (OTS)/hexane solution to make the entirety of the paper hydrophobic. Hydrophilic channels and detection zones could then be patterned into the hydrophobic paper by exposing it to deep UV and ozone while using a patterned mask (photolithography). Glavan et al. [38] chose a highly physical approach to creating microchannels by using an electronic craft-cutting/engraving tool to carve microchannels into cardstock paper. Carved microchannels could be configured via engraving tips of varying dimensions, and the craft cutter could be controlled to manufacture microchannel widths of 45–300 μm and depths of 50–300 μm. After microchannel creation, the patterned cardstock paper in these experiments was silanized with alkyl or fluoroalkyl trichlorosilane to make it hydrophobic or omniphobic, respectively. Completed μPADs could sustain pressure-driven flow experiments in open microchannels.

Schilling et al. [39] used layers of wax-patterned paper and laser-printed toner films to simultaneously create hydrophobic barriers and thermally bond them together to create 3D μPADs for multiple diagnostic assays.

Nie et al. [40] used a commercial CO_2 laser cutting/engraving machine in a one-step fabrication process that allowed for the creation of hollow microfluidic structures. Chitnis et al. [41] also used a CO_2 laser to fabricate μPADs, but in a different manner. They created hydrophilic channels by using the laser to selectively cut paths in the surface material of surface-treated hydrophobic papers (parchment, wax, and palette papers) by optimizing both laser strength and cutting/scanning speed.

3.2.3 Detection Methods

Ongoing research is highly active in areas that could enhance and improve μPAD diagnostic detection methods. Among the methods garnering most attention recently, electrochemical, electrochemiluminescence, and surface-enhanced Raman spectroscopy approaches are prominent, but traditional colorimetric detection methods continue to be used and developed. The subsections below describe recent advances in detection techniques used in paper-based diagnostic devices.

3.2.3.1 Colorimetric Detection Method

Colorimetric detection, a method based on the detectable color reaction between analytes and colorimetric reagents, is common and simple for use with μPADs [3, 5]. In the period covered by this review, some authors have made remarkable advances in μPAD-based colorimetric detection for clinical assays. Ornatska et al. [42], for instance, have used colorimetry to detect glucose by employing ceria nanoparticles and glucose oxidase immobilized on paper following silanization with aminopropyltrimethoxysilane (APTS). In this research, a chitosan layer was also employed for glucose oxidase stabilization. This ceria-based colorimetric detection method was capable of detecting glucose with a limit of detection (LOD) of 0.5 mM. Peng et al. [43] took advantage of the interaction of alkaline phosphatase (ALP) with colloidal calcium carbonate to increase the retention of ALP in filter paper following filtration as a means of pre-concentrating. They attained a very low LOD (1.1 nM) for paper-based colorimetric assay of ALP with a p-nitrophenyl phosphate substrate. To achieve colorimetric detection of cysteine, Tseng et al. [44] used nanotechnology, specifically a photothermal process to produce metallic (gold or silver) nanoparticles on paper. They employed a sputtering method and a KrF excimer laser to subsequently melt the metal film, which was then cooled so that adhered metal nanoparticles were generated on their paper substrate due to surface tension kinetics.

3.2.3.2 Fluorescence Detection Method

Not as practical in the field because it requires external instrumentation, fluorescence detection in μPADs has been shown to provide good analytical performance [11]. Yuan et al. [45] were able to detect glucose and catechol in a μPAD via fluorescence detection. They used a hybrid material synthetized by encapsulating

CdTe quantum dots and enzymes within a film of poly(diallyldimethylammonium chloride). Using a fluorescence quenching effect and hybrid material containing glucose oxidase and tyrosinase, they showed that glucose and catechol, respectively, could be detected. Some researchers discovered methods by which they could pretreat paper with diagnostic analytes so that target areas for specific assays could be predesigned into paper-based diagnostic tools. Research by Yu et al. [46] demonstrated that cellulose paper could be modified with divinyl sulfone so that carbohydrates, proteins, and DNA could be covalently immobilized on paper for colorimetric and fluorometric bioassays. Liang et al. [47] showed that polystyrene microbeads functionalized with antibody for goat immunoglobulin G (IgG) could be immobilized on paper. To detect IgG, Liang and colleagues used cyanine dye (Cy3) and gold nanoparticles (AuNPs) conjugated with secondary antibody (anti-goat IgG) for sandwich-type fluorescence and colorimetric immunoassays, respectively. Electing a non-paper-focused approach, Yildiz et al. [48] fabricated a poly(vinylidene fluoride) (PVDF) porous membrane that they then modified with poly(3-alkoxy-4-methylthiophene) to fluorometrically detect a microRNA sequence associated with lung cancer. Noor et al. [49] used imidazole ligands to immobilize quantum dots on paper and then used the immobilized quantum dots as donors for fluorescence resonance energy transfer (FRET). Ultimately, by using this paper-based FRET method, they were able to detect nucleic acid hybridization with an LOD of 300 fmol.

3.2.3.3 Chemiluminescence Detection Method

Comparable to fluorescence methods in μPADs, chemiluminescence detection relies on the capacity to measure light emitted during chemical reactions. Yu et al. [50] relied upon the measurable chemiluminescence of hydrogen peroxide (produced) and a rhodamine derivative to develop a multiplexed μPAD that could simultaneously determine the presence of glucose and uric acid by respective reactions with glucose oxidase and uricase (urate oxidase). Using a chemiluminescent ELISA approach, Wang et al. [51] employed a wax screen-printing method [29] and μPADs pre-coated with chitosan to covalently immobilize antibodies for α-fetoprotein (AFP), cancer antigen 125 (CA125), and carcinoembryonic antigen (CEA). Using these wax screen-printed μPADs, Wang and colleagues were able to perform a 4-iodophenol-enhanced luminol chemiluminescence assay to detect AFP, CA125, and CEA in human serum. In order to create a μPAD for CEA detection in human serum, Wang et al. [52] used a periodate oxidation reaction to generate aldehyde groups on cellulose paper for covalent bonding of antibodies, completing their detection technique with a chemiluminescence immunoassay. In further chemiluminescence research using paper, Wang et al. [53] successfully used a DNA biosensor employing N,N′-disuccinimidyl carbonate to immobilize captured DNA strands on paper and labeled carbon dots of nanoporous gold to improve detection sensitivity.

3.2.3.4 Electrochemiluminescence Detection Method

Electrochemiluminescence (ECL) approaches use a set of electrodes to trigger and control a chemiluminescence reaction involving a luminophore compound that is excitable under electrical stimulation [54]. This approach benefits from combined methodology; it blends electrochemical and luminescence techniques that impart both good selectivity and good sensitivity for detection in μPADs. Other luminophore reagents have been used such as those demonstrated in the research published by Delaney et al. [55] who performed an ECL detection in μPADs using tris(2,2'bipyridyl)ruthenium(II) ($Ru(bpy)_3^{2+}$) and inkjet-printed μPADs (see Fig. 3.2) that were laminated with screen-printed electrodes for ECL detection. ECL emission was detected using a photodetector and a mobile phone camera. Shi et al. [56] improved the reproducibility of ECL emissions for H_2O_2 solutions by immobilizing and stabilizing CdS quantum dots on a double-sided carbon adhesive tape supported by an indium tin oxide (ITO) glass that acted as a working electrode for ECL in μPADs. Ge et al. [57] produced a multiplexed ECL immunoassay detection device by incorporating eight screen-printed carbon working electrodes in a 3D μPAD. In their ECL immunoassay approach, Yan et al. [58] also developed a 3D μPAD, but used chitosan and glutaraldehyde cross-linking to immobilize CEA, a cancer biomarker, in a screen-printed carbon electrode.

Wang et al. [59] improved upon the commonly used potentiostat system by employing a 3-V lithium battery and a simple electronic circuit that allowed them to specifically tune voltage to elicit an ECL immunoassay in μPADs. Li et al. [60] also used a battery-trigged ECL system, but employed functionalized, nanoporous silver (NPS) to amplify the signal and improve sensitivity for detecting prostate-specific antigen (PSA) tumor markers and CEA. In an innovative use of materials, Xu et al. [61, 62] employed screen-printed carbon working electrodes modified with composite films of poly(sodium 4-styrenesulfonate)-functionalized graphene/Nafion [61] and Fe_3O_4 nanocrystal clusters/graphene sheets in order to improve the

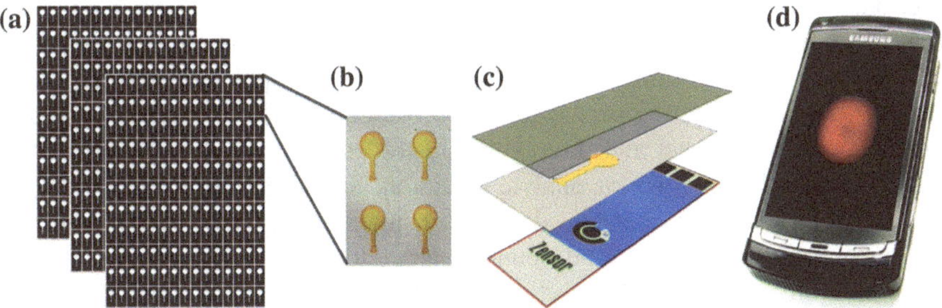

Fig. 3.2 Fabrication of a μPAD with ECL detection: **a** sheet of paper patterned with many microfluidic devices; **b** paper devices modified with luminophore reagent Ru(bpy)32+; **c** paper microfluidic device is aligned to the face of a screen-printed electrode and laminated with transparent film; **d** a camera phone is used to capture and analyze the ECL emission. Reproduced with permission from Delaney et al. [55]

immobilization and electron transfer of the luminophore agent $Ru(bpy)_3^{2+}$ and obtain highly sensitive solid-state ECL detection in μPADs [62]. Yan et al. [63] developed a 3D μPAD (origami technique) with oligonucleotides (aptamers) immobilized on a porous working electrode in order to increase ECL detection sensitivity.

3.2.3.5 Electrochemical Detection Method

For several reasons, electrochemical detection has shown that it is highly suitable and useful in μPADs [64–67]: (i) It is easily miniaturized; (ii) it can provide high sensitivity; and (iii) it is more rugged than optical detection methods. In the period covered by this review, implementation, electrode modification, and multiplex electrochemical detection advances have been reported. Liu and Crooks [68] described their development of a μPAD that, essentially, has an onboard electrochromic display feature. They used an electrochemical sensor powered by an integrated metal/air battery, with clinically sampled urine as their electrolyte and transparent ITO electrodes in both the sensor and the battery assembly. Detection of glucose and H_2O_2 was via colorimetric change of a Prussian blue spot on their ITO electrode. Rattanarat et al. [69] modified the surface of a three-layered μPAD, assembled with commercial, disposable electrodes, by treating it with sodium dodecyl sulfate (SDS) to pre-concentrate and increase selectivity for dopamine.

Godino et al. [70] developed a disposable hybrid paper/polymer microfluidic device using chromatography paper, poly(methylmethacrylate) (PMMA) film, and an electrochemical cell. Hydrophobic features and electrodes were deposited on the chromatographic paper via the combined use of wax printing and a patterned adhesive stencil, and a pressure-sensitive adhesive was then used to bind the assembly to a patterned PMMA film. This paper/PMMA microdevice was electrochemically characterized by cyclic voltammetry using 1 mM ferrocyanide. Santhiago et al. [71] electrochemically quantified cysteine using a μPAD and carbon paste electrodes modified with cobalt phthalocyanine (catalyst). The electrodes were fabricated by applying carbon paste on paper using a polyester (transparency film) stencil that was generated by laser engraving. Shiroma et al. [72] described a μPAD that was capable of chromatographically separating and electrochemically detecting paracetamol and 4-aminophenol. Comprising a hydrophilic separation channel and three gold electrodes deposited by sputtering, this μPAD employed amperometric analyte detection. Santhiago and Kubota [73] created a μPAD for electrochemical detection using inexpensive pencil graphite as external working, reference, and counter electrodes that were screen-printed on paper with silver ink. Noiphung et al. [74] developed a dual-membrane electrochemical μPAD to separate plasma from whole blood and determine glucose concentration in human whole blood. The dual-membrane assembly allowed for the isolation of plasma, which then flowed to a detection zone containing immobilized glucose oxidase. Enzymatically produced hydrogen peroxidase was detected using screen-printed electrodes modified with Prussian blue. The results of these experiments were in good agreement with the spectrophotometric reference.

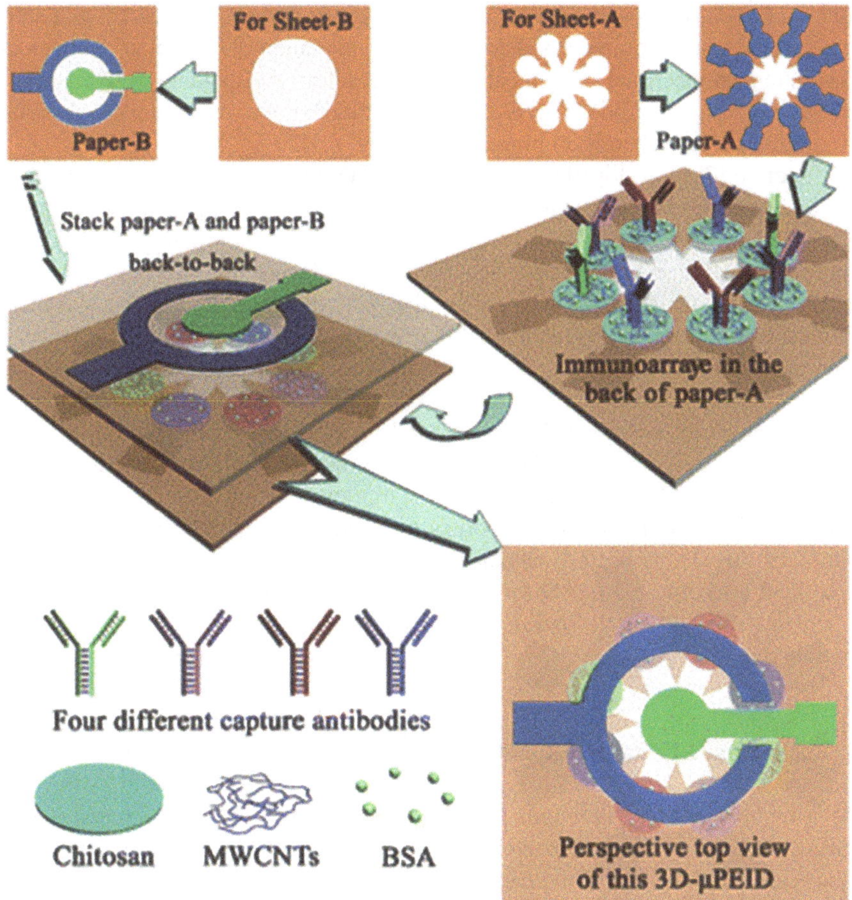

Fig. 3.3 Schematic diagram of a 3D μPAD fabrication process with eight working electrodes for electrochemical immunoassay for the detection of tumor markers. Reproduced with permission from Zang et al. [78]

Lu et al. [75] created a three-electrode system in which electropolymerization of the working electrode permitted pH and glucose determination on coated paper. Electrodes were fabricated using an inkjet printing technique. Employing an array of addressable screen-printed electrodes, Ge et al. [76] developed a μPAD to amperometrically detect tumor markers. The capture and immobilization of antibodies on paper were carried out by using carbon nanotubes and chitosan, and a secondary antibody to horseradish peroxidase was coupled to AuNPs/carbon nanotubes in a composite for use as an electrochemical probe.

Carbon nanotubes (multiwalled) were also used by Wang et al. [77] to enhance electronic conductivity, and hence electrochemical immunodetection, in a 3D μPAD. By pre-coating their paper substrate with chitosan and cross-linking with glutaraldehyde, they were able to immobilize antibodies for both carcinoma antigen and CEA so that tumor markers could be simultaneously detected. Zang et al. [78] extended electrode implementation methodology by including eight screen-printed working electrodes (PWEs) in a 3D μPAD that was capable of performing multiplexed electrochemical immunoassays (Fig. 3.3).

Liu et al. [79] developed an origami-based 3D μPAD [34] as a means of creating a concentration electrochemical cell to detect adenosine. High selectivity for electrochemical detection was facilitated by immobilizing aptamers for adenosine in polystyrene microbeads. Sensitivity was cleverly enhanced by using an external capacitor that was charged by the potential generated within the electrochemical cell itself. The analytical signal was the instantaneous current when the capacitor discharged. Li et al. [80] developed an origami-based 3D μPAD to detect cancer biomarkers (CEA and AFP) using a multiplexed electrochemical immunoassay approach. In this research, a NPS layer was applied to paper fibers placed over a screen-PWE. Capture antibodies were then applied to the assembled NPS-PWE, and a secondary antibody coupled to a nanocomposite containing a combination of nanoporous gold, chitosan, and absorbed metal ions (Cu^{2+} or Pb^{2+}) was used to enhance the amperometric response. Lu et al. [75] described a folding 3D μPAD for electrochemical DNA detection that employed a screen-PWE and single-stranded capture DNA attached to an AuNP/graphene composite.

3.2.3.6 Surface-Enhanced Raman Spectroscopy Detection Method

As a highly sensitive detection method, surface-enhanced Raman spectroscopy (SERS) is an attractive alternative. In most cases, paper is spotted with metal nanoparticles and the pores of the paper act as reservoirs to increase SERS nanoparticle concentration in selected detection zones. To detect Rhodamine 6G, malathion, heroin, and cocaine, Yu and White [81, 82] created a highly sensitive paper-based SERS microdevice using an inkjet printing technique to deposit silver nanoparticles (Fig. 3.4) on μPADs. Ngo et al. [83] added AuNPs to a paper substrate via a dipping method and employed 4-aminothiophenol (4-ATP) as a Raman probe to monitor the adsorption of AuNPs onto cellulose fibers. They also investigated the concentration influence of nanoparticles and 4-ATP on the SERS signal. Chen et al. [84] modified filter paper with poly(vinyl pyrrolidone) (PVP) and silver colloid and then used magnetic beads to perform an immunoassay for mouse IgG. Abbas et al. [85] developed a highly sensitive μPAD that could separate, preconcentrate, and detect analytes at a subattomolar level. The authors attribute the sensitivity of their device to the addition of gold nanorods for SERS detection (dipping method), the novel cutting of their μPAD substrate paper into a starlike shape, and their choice to back their substrate with a polyelectrolyte coating to enhance capillary action and improve separation and concentration of analytes.

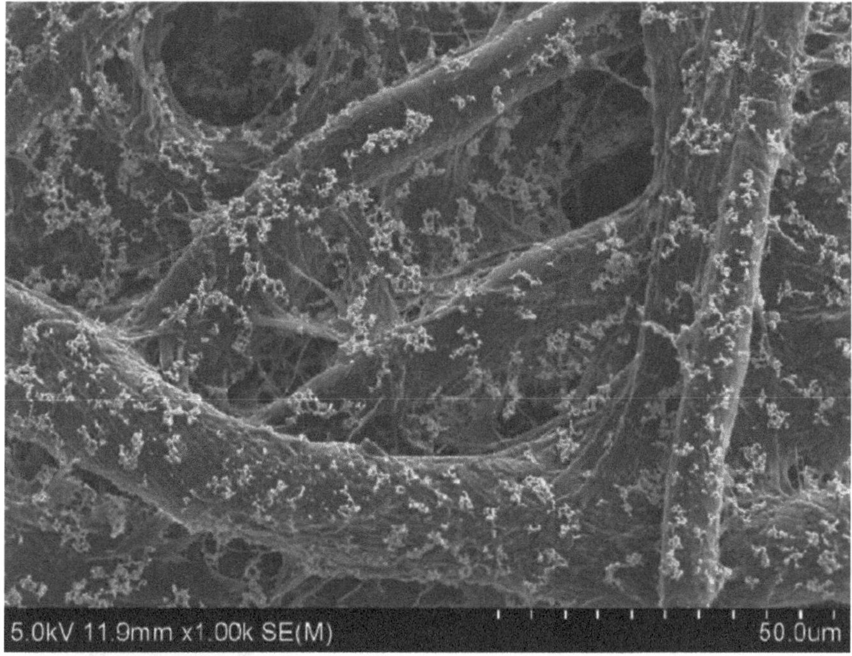

Fig. 3.4 Scanning electron microscopy (SEM) image of cellulose paper modified with silver nanoparticles for SERS detection. Adapted with permission from Yu and White [81]

3.2.3.7 Other Detection Methods

This subsection describes some reported detection methods used in μPADs that employed approaches that do not fit well into any of the previously described detection method groups. Lewis et al. [86], for instance, used wax-patterned paper arranged in stacked layers to create 3D μPADs capable of detecting hydrogen peroxide with two different readout mechanisms: (1) sample flow time as it moved through the μPAD and (2) the number of colored reaction zone bars produced by the assay. When hydrogen peroxide reacted with a hydrophobic compound deposited within the paper, hydrophilic spaces were produced that changed the wettability of the paper and improved sample wicking. Greater hydrogen peroxide solution concentration hastened flow-through time and increased the number of colored reaction zones as measured in bar format (Fig. 3.5).

Tian et al. [87] produced a bioplasmonic paper substrate by immobilizing gold nanorods functionalized with antibodies to detect a kidney cancer biomarker (aquaporin-1 protein) in artificial urine. Their bioplasmonic paper acted as a localized surface plasmon resonance (LPSR) substrate with an LOD of about 0.16 pM. Ge et al. [88] described a chemiluminescence-based μPAD that converted emitted

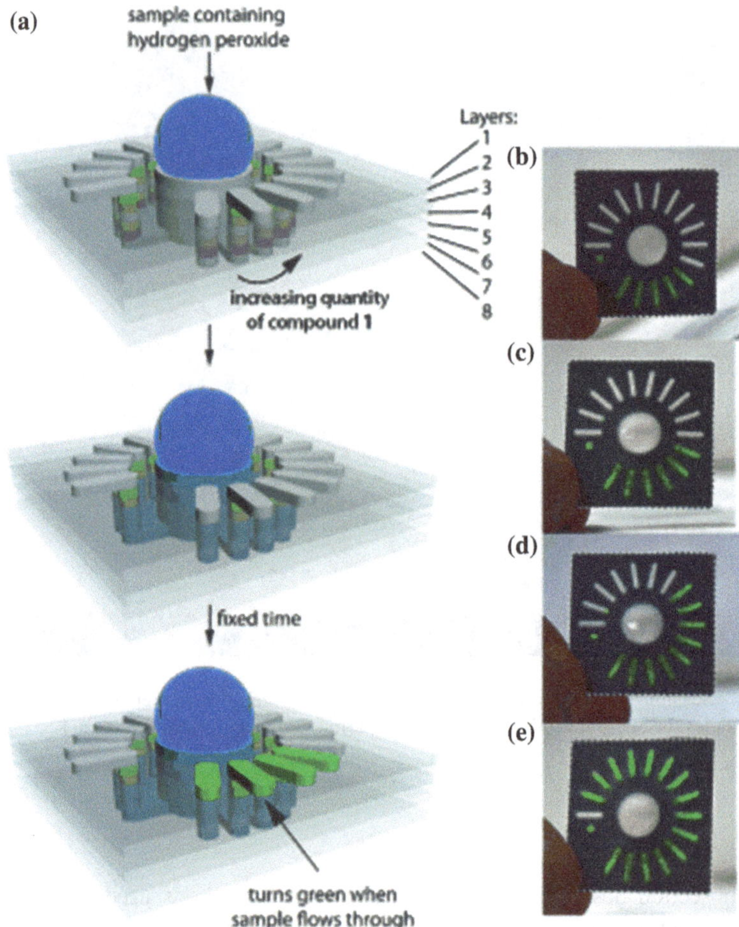

Fig. 3.5 Diagram of 3D μPAD and the assay procedure with readout using number of produced colored bars (test zones): **a** representation of the liquid flow in the 3D μPAD; **b–e** readout after 10-min assay with hydrogen peroxide solution at **b** 1 mM, **c** 35 mM, **d** 75 mM, and **e** 100 mM. Reproduced with permission from Lewis et al. [86]

light into an electric current (photocurrent) that charged an integrated paper supercapacitor. The instantaneously discharging current from the capacitor acted as the analytical signal. Chemiluminescence was achieved using N-(aminobutyl)-N-(ethylisoluminol)-functionalized AuNPs and an adenosine triphosphate (ATP) aptamer for molecular recognition of ATP. Wang et al. [89] described a multiplexed μPAD with a similar photoelectrochemical method that was capable of detecting four different cancer biomarkers in human serum.

3.2.4 New Functions and Design

A host of notable advances in μPAD capabilities have been achievable following insights into mixing, flow control and flow dimension, and the capacity of μPADs to self-power. In regard to flow dimension, Fu et al. [90] developed a platform that was adaptable to a conventional paper-based lateral flow test but could perform assays in two separate dimensions. This added dimensional capacity can be used to enhance assay sensitivity and allows the μPAD to withstand rinse and signal amplification steps amenable to a commercial strip test to detect chorionic gonadotropin as in a pregnancy test.

Leveraging centrifugal forces, Hwang et al. [91] adapted a paper strip and a centrifugal microfluidic platform to show how such forces can actively control both flow rate and invert the flow direction in paper. Lutz et al. [92] developed a 2D μPAD with the capacity to accept a programmable assay sequence that controlled flow arrival and duration for all reaction zone reagents (Fig. 3.6). Capable of performing complex, automated, multistep chemical assays, the authors leveraged the unique capacity of this 2D μPAD to perform experimental, theoretical, and computer simulation studies to investigate flow behavior in their multistep μPAD [93].

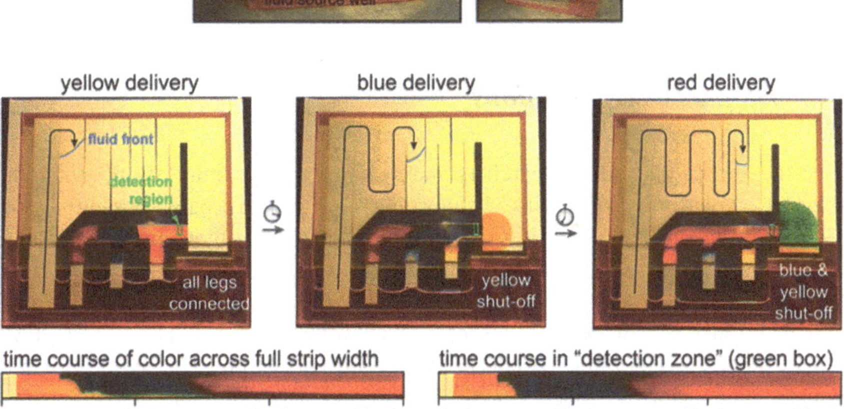

Fig. 3.6 Two-dimensional paper networks with automated flow delivery sequence for controlling the arrival time and flow duration of each reagent in the reaction zones. Adapted with permission from Lutz et al. [92]

By laser printing layers of toner on both sides of a μPAD, Schilling et al. [94] were able to completely enclose their established microchannels. This improved fluid wicking rate, reduced total evaporation, and helped avoid solution contamination. Kwong and Gupta [95] combined μPAD operations, such as analyte separation and fluid manipulation on paper, by depositing functional polymers, i.e., poly(methacrylic acid), poly(dimethylaminoethyl methacrylate), and poly(o-nitrobenzyl methacrylate), onto their paper substrate using initiated chemical vapor deposition (iCVD) methodology. Poly(methacrylic acid) and poly(dimethylaminoethyl methacrylate) acted as ion-exchange coatings to separate analytes, and poly(o-nitrobenzyl methacrylate) acted as a UV-responsive switch on their paper substrate. Jahanshahi-Anbuhi et al. [96] demonstrated an approach to accelerate liquid flow and improve assay speed by sandwiching paper between flexible polyester films, which improved liquid wicking speed by one order of magnitude as a result of negative capillary pressure in the liquid. Rezk et al. [97] showed that sound, specifically surface acoustic waves (SAW) at 30 MHz, enhanced mixing and induced more uniform, reproducible, and faster mixing of solutions on a μPAD than mixing via passive capillary-driven flow.

Thom et al. [98] employed integrated galvanic cells as "fluidic batteries" on a multilayer μPAD to generate power following sample addition. These paper-integrated galvanic cells were assembled with electrodes of Ag and Al, electrolytes ($AgNO_3$ and $AlCl_3$), and salt bridges containing $NaNO_3$. Using a serial and parallel association of galvanic cells, this paper-based battery was able to power an LED for 8.2 min. In another article, Thom et al. [99] investigated several related influential parameters regarding his integrated galvanic array including layout configuration, electrolyte concentration, electrode material and size, electrical resistance of salt bridge, and the number of galvanic cells in serial and parallel association that could be controlled to obtain predictable and tunable fluidic batteries on a paper substrate.

3.2.5 Diagnostic Applications

Over the recent decades, researchers have shown greater interest in examining paper-based platforms for in vitro diagnostic assays. Already, paper-based diagnostics, the most translationally achievable application for this biotechnology, is used to examine clinical samples such as blood, saliva, tears, aqueous humor, seminal fluid, and more [1, 14, 16, 17, 25, 64, 100, 101]. Clearly versatile, such platforms offer additional advantage in terms of sample volume needed. Developed μPAD devices are already portable, small, and compact, but the physical nature of paper provides them with the added benefit of requiring only minimal sample and reagent volumes. As previously shown, diagnostic paper-based platforms can be easily integrated with several analytical detection approaches including those most common, i.e., colorimetric detection, fluorescence detection, chemiluminescence detection, electrochemical detection, and transmittance detection [102]. Further, research provides evidence that paper-based platforms are comparably accurate, albeit considerably less costly, to existing plastic-based microplate platforms. While paper platforms have some hurdles that require being overcame such as more greater integration with existing instrumentation, the future of such approaches is scientifically, environmentally, and economically promising. Here, we shall describe advances in paper-based assay platforms designed with a focus on clinical diagnostics.

3.2.5.1 Paper-Based Metabolic Assays

Here, we shall define metabolic assays as those that detect the presence or levels of specific metabolic analytes such as glucose, cholesterol, or aminotransferase in blood. Detection has a diagnostic purpose and a target therapy goal, but assay results need to be translatable using efficient, rapid, robust, and inexpensive platforms. Traditionally, high-throughput assays have required the use of plastic microplates, which incur high cost not only in microplate plastics, but also in the quantity of costly reagents (e.g., mAbs, enzymes, substrates) required. Because paper is less expensive than microplate plastics, and paper platforms require less reagent volume, their use incurs much less cost. The detection of metabolic analytes toward a diagnostic/translational medicine goal has been successfully carried out using paper-based platforms. Using enzymatic reaction chemistry or small-molecule dyes and colorimetric assays, paper has been used to monitor glucose, BSA, nitrites, ketones, ALP, aspartate aminotransferase–alanine aminotransferase (AST/ALT), and cholesterol (Table 3.1) [3, 64]. Martinez et al. successfully detected both glucose and BSA using a paper-based platform with LODs of 2.5 mM and 0.38 μM, respectively (Fig. 3.7a) [9]. Using a paper-based platform, Klasner et al. were able to detect nitrites and ketones (acetoacetate) with LODs of 5 μM and 0.5 mM, respectively (Fig. 3.7b) [103]. Paper-based detection of AST and ALT in clinical serum samples has been demonstrated and validated with greater than 90 % accuracy (Fig. 3.7c) [17]. Vella et al. developed a paper-based diagnostic tool that not only integrated sample preparation and metabolic assay, but also measured two liver function markers, ALP and AST, as well as total serum protein level (Table 3.1) [16]. Wong et al. demonstrated that a simple kitchen utensil, an eggbeater, could be used as a centrifuge to separate plasma from human blood samples in resource-limited settings. The goal of Wong and colleagues was to detect cholesterol in human serum by using a cholesterol oxidase method that, in a reaction with Amplex® Red, resulted in a detectable pink colorimetric reaction on their paper-based system. This method allowed for the detection of cholesterol at levels as low as 0.67 mM, with a limit of quantitation (LOQ) of 0.8 mM, which was acceptable for commercially available whole blood >5.2 mM (Fig. 3.7d) [104]. Matsuura et al. successfully demonstrated a paper-based fertility-analyzing device for sperm that was robust and easy to use. They patterned a piece of paper and, using small-molecule dye-based bioanalysis, measured the activity of mitochondrial dehydrogenase enzyme with yellowish 3-(4,5-dimethyl thiazol-2-yl)-2,5-diphenyl tetrazolium salt (MTT). Such a device could be used "in the field" to

Table 3.1 Clinically related biomarkers in paper-based platform (Adapted from Ref. [116])

Biomarker	Fields in paper-based platform	Disease/condition	Sample	Application	µPAD dynamic range	LOD	Cutoff value/reference in clinics	Reference
ALT (alanine transaminase)	Metabolic assay	Liver injury	Serum	Point-of-care diagnostics	0–400 U/L	53 U/L	>40 U/L	[17]
AST		Liver injury	Serum			84 U/L; 44 U/L	>40 U/L	[16, 17]
(aspartate transaminase)		Liver injury	Serum		44–200 U/L	44 U/L	30–120 U/L	[16]
ALP (alkaline phosphatase)		Biliary obstruction/ hypophosphatemia	Serum		15–1000 U/L	15 U/L	30–120 U/L	[16]
Cholesterol		Metabolite disease	Plasma		0.8–6.5 mM	0.67 mM	>5.2 mM	[104]
Nitrite		End-stage disease	Salivary (artificial)		5–2000 µM	5 µM	1–40 nM (normal); 40–160 µM (patients)	[103]
Ketone		Diabetic ketoacidosis/ end-stage disease	Urine (artificial)		5–16 mM	0.5 mM	$\leq$1.9 mM for "low," between 2.9 and 3.9 mM for "moderate," and $\geq$ 7.8 mM for "high"	[103]
Glucose		Renal diseases/metabolite disease	Urine (artificial)		0–20 mM/3–50 mM	2.5 mM	>0.8 mM	[3, 103]
Lactate		Metabolite disease	Serum		0–50 mM	0.36 ± 0.03 mM	>3.5 mM	[64]
Uric acid		Metabolite disease	Serum		0–35 mM	1.38 ± 0.13 mM	Abnormal range: > 0.4 mM; < 0.1 mM	[64]
Albumin		Renal diseases/metabolite disease	Urine (artificial)		0.38–60 µM; 0.38–7.5 µM	0.38 µM	Clinical albuminuria ([protein] > 0.1 mM)	[3, 16]

(continued)

Table 3.1 (contined)

Biomarker	Fields in paper-based platform	Disease/condition	Sample	Application	µPAD dynamic range	LOD	Cutoff value/reference in clinics	Reference
AFP (alpha-fetoprotein)	P-ELISA	Hepatocellular carcinoma	Serum	Clinically based diagnostics	0.1–35 ng/mL	0.06 ng/mL	>25 ng/mL	[51]
CA125 (cancer antigen L25)		Ovarian cancer	Serum		0.5–80 U/mL	0.33 U/mL	>35 U/mL	[51]
CEA (carcinoembryonic antigen)		Colorectal cancer	Serum		0.1–70.0 ng/mL	0.05 ng/mL	>5 ng/mL	[51]
PSA (prostate-specific antigen)		Prostate cancer	Serum		0.5–50 µg/L	360.2 ng/L	>4 µg/L	[32]
NC16A		Bullous pemphigoid	Serum or blister fluid		1–50 µg/mL			[101]
NPY (neuropeptide Y)		Post-traumatic stress disorder (PTSD)	Saliva		10 pM–100 nM	1 pM	50–100 pM (normal), 400–1400 pM (stress)	[106]
hCG		Pregnancy test and trophoblastic disease	Urine		1–500 ng/mL	4 ng/mL (40 mIU/mL)	<5.0 mIU/mL (non-pregnant women)	[107]
VEGF (vascular endothelial growth factor)		Retinal ischemic	Aqueous humor		$1O^{14}$–10^{-6} g/mL	10^{-14} g/mL	14.4 ± 8.5 pg/mL (normal), PDR (740.1 ± 267.7 pg/mL), AMD (383 ± 155.5 pg/mL), and RVO (219.4 ± 92.1 pg/mL)	[25]
Lactoferrin		Disorders of the corneal epithelium	Tear		0.5–3 mg/mL	0.3 mg/mL		[100]
Serotype-2 dengue		Infectious disease	Serum		100 pg/mL–14 µg/mL	100 pg/mL	4 ng/mL (conventional ELISA)	[19, 26]
HBV		Infectious disease	Plasma		10–300 IU/mL	50 IU/mL	47 IU/mL(TaqMan fluorescence technology)	[108]
HCV		Infectious disease	Serum		26.7 fmol–257 amol	6.7 amol (0.1 pg)		[110]

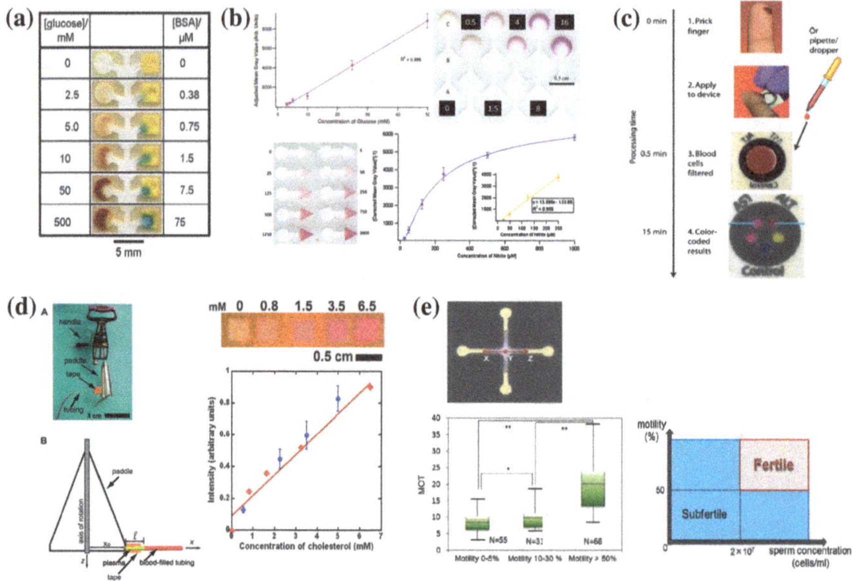

Fig. 3.7 Paper-based metabolic assays. Paper-based microfluidic devices (µPADs) are made from paper patterned with hydrophilic channels and bounded by hydrophobic barriers and are designed to save time and cut cost. **a** Martinez et al. [3] reported successful detection of glucose and bovine serum albumin (BSA) in artificial urine using a paper-based platform with a limit of detection (LOD) of 2.5 mM for glucose and 0.38 µM for BSA [9]. **b** Klasner et al. [103] used µPADs to monitor urinary ketones, glucose, and salivary nitrite levels in urine. Many markers are related to metabolic syndromes, e.g., failure to regulate glucose can result in hypoglycemia, diabetic ketoacidosis (DKA), and diabetic coma. In these studies, the LOD for acetoacetate in artificial urine was 0.5 mM, while the LOD for salivary nitrite was 5 µM, and the time of device fabrication to the time of test results was 25 min. The assay for urinary ketones demonstrated, for the first time, the ability of these devices to perform online chemical derivatization. This assay used a two-part format that allowed chemical modification prior to detection. In this assay, the acetoacetate in urine first reacts with glycine to form an imine derivative [103]. **c** Pollock et al. [17] demonstrated a µPAD method to be used in resource-limited settings or developing countries that could provide POC monitoring of drug-related hepatotoxicity for individuals on standard therapies for tuberculosis (TB) and/or HIV. This method was less expensive and more rapid (<15 min) than traditional measures in providing visual measurements of AST and ALT in whole blood or serum [17]. **d** Wong et al. [104] developed a process that employed a hand-powered eggbeater to overcome disadvantages of a standard centrifuge, which requires a source of electrical power, and can be bulky, difficult to repair, and expensive (>$400). The device-isolated plasma was more easily used to run a cholesterol assay on paper. For resource-limited areas, other infectious diseases such as HBV and cysticercosis can also, in principal, be diagnosed in a paper-based POC assay [104]. **e** Using a paper-based concept to evaluate fertilization capacity using various types of low-cost, easy-to-use, and rapid devices, Matsuura et al. (2014) reported an inexpensive but robust and easy-to-handle device for monitoring the health status of human sperm made by patterning a piece of paper and measuring the activity of a specific enzyme that could readily be reference to standard WHO values for evaluating sperm concentration ($>2 \times 10^7$) and motility (>50 %) [105]

evaluate fertility levels without the need to consult doctors. Results could be easily compared with WHO reverence values for healthy sperm concentration (>2×10^7) and motility (>50 %). Moreover, the duration and cost of one complete test are only 30 min and $0.03, respectively (Fig. 3.7e) [105]. These results in mind, mere paper may be well poised to provide a suitable replacement for currently expensive, wasteful, and complex microplate-based analyses.

3.2.5.2 Paper-Based ELISA

An enduring standard, ELISA is still widely used in the selection of therapeutic antibodies and to monitor the effect on virus titer or disease-associated biomolecules following therapeutic drug delivery. The methodology employs an antibody-specific signal amplification process facilitated by conjugation with high-turnover catalytic enzymes and an enzymatic substrate that produces a detectable color signal.

The first account of paper-based ELISA (P-ELISA) was made by Cheng et al. [20] in a research article describing the successful detection of IgG and human immunodeficiency virus (HIV) antigen titer via colorimetric assay. Exceptionally efficient, P-ELISA drastically reduces the need for reagent volumes to as low as one twenty-fifth the volume needed for current microtiter plate processes. Further, it reduces reaction time to one fifth of the time required for conventional ELISA (Fig. 3.8a) [20]. In addition to detecting IgG and HIV, P-ELISA has been used to detect VEGF in extraordinarily minute clinical aqueous humor samples. In ophthalmological experiments by Hsu et al. [25] using samples from thirteen patients ($N = 13$), with senile cataract as the control, they found that the mean aqueous VEGF level from patients with proliferative diabetic retinopathy ($N = 14$), age-related macular degeneration ($N = 17$), and retinal vein occlusion ($N = 10$) showed that VEGF increases to 740.1 pg/mL, 383 pg/mL, and 219.4 pg/mL, respectively (LOD as low as 14.4 pg/mL) (Fig. 3.8b) [25]. In regard to cancer-related studies, Wang et al. found that they could perform chemiluminescence ELISA by using chitosan-modified paper so that a linear range of 0.1–35.0 ng/mL for α-fetoprotein (AFP), 0.5–80.0 U/mL for cancer antigen 125 (CA-125), and 0.1–70.0 ng/mL for CEA could be achieved (Table 3.1) [51]. Nie et al. innovatively developed a rapid patterning method using porous pens, rather than the more complex process of printing permanent hydrophobic symbols on paper, and discovered that their process could be used to colorimetrically detect and quantify a prostate cancer marker, PSA. Based on dot-immunogold staining assays coupled with gold enhancement amplification, the dynamic range of PSA concentrations was determined to range from 0.5 to 50 ng/mL, and the LOQ of PSA was determined to be 360.2 pg/mL. These detection values cover cutoff values for the four tumor markers in clinical diagnoses (Table 3.1) [32]. Hsu et al. pioneered the application of P-ELISA to detect autoimmune antibodies in human specimen samples of serum and blister fluid. Using only 2 μL of serum or blister fluid and taking a mere 70 min, Hsu and colleagues were able

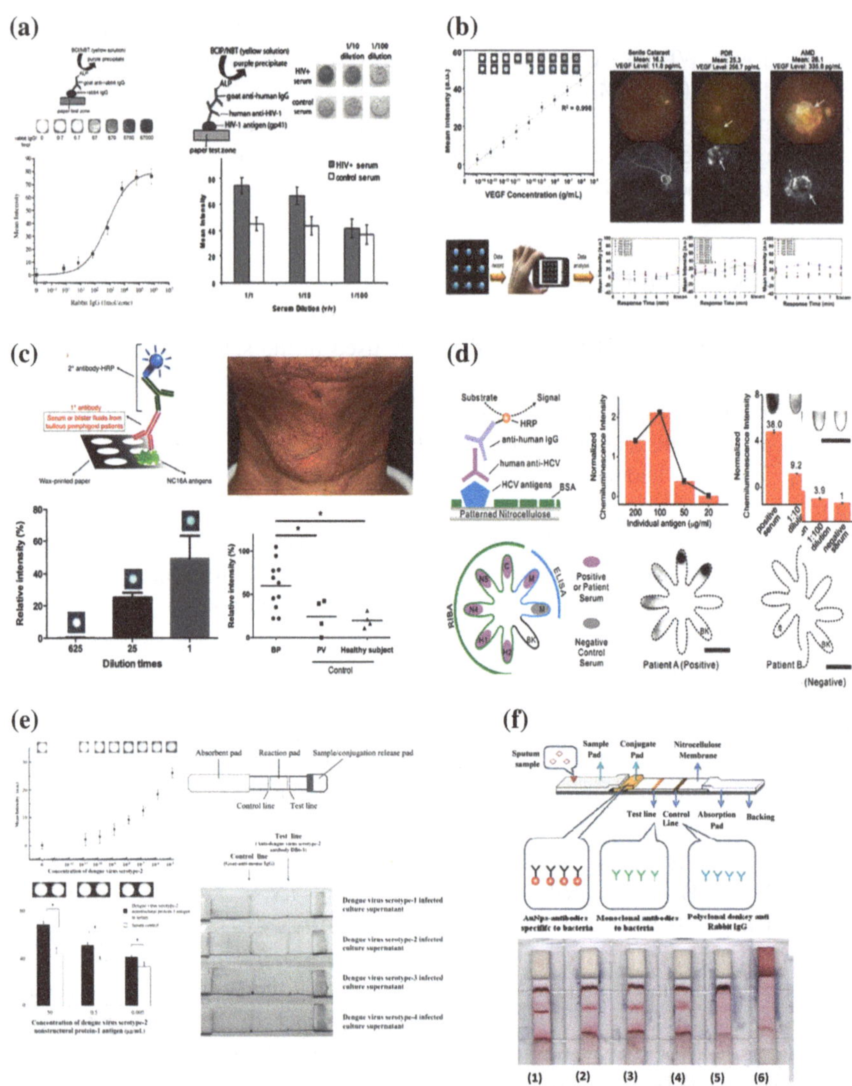

to detect anti-NC16A autoimmune antibodies (indicative of bullous pemphigoid or BP), thus offering a significantly advantageous tool for point-of-care diagnostics for afflicted patients (Fig. 3.8c) [101]. Yamada et al. reported on an intriguing fluorescence P-ELISA method that relied on clinical samples of human tears to detect lactoferrin. Although the LOD was much higher compared to that reported for the ELISA kit (1 ng/mL), this method provides a promising diagnostic alternative for detecting lactoferrin that is eminently usable for point-of-care diagnostics. The results of Yamada's experiments showed that assays can be completed within 15 min, requiring only the application of a fresh tear sample to the sampling area

◀ **Fig. 3.8** Paper-based ELISA. **a** P-ELISA was first proven by Cheng et al. [20] to quantitate for rabbit IgG. Following this success, a method was demonstrated for the detection of antibodies to the HIV-1 envelope antigen gp41 using P-ELISA that was efficient, rapid, and inexpensive [20]. **b** Hsu et al. [25] developed a P-ELISA that used an indicator, VEGF, to monitor eye disease in patients with retinal ischemia. For clinical detection, this diagnostic method required only 2 μL of aqueous humor. The mean aqueous VEGF level from thirteen patients with senile cataract as the control was 14.4 pg/mL. The mean aqueous VEGF level from other patients with age-related macular degeneration (AMD) showed VEGF increases of 383 pg/mL. Thus was created an inexpensive and minimally invasive diagnostic approach that demonstrated high sensitivity and short operation duration and required only very small clinical sample volume [25]. **c** Hsu et al. [101] demonstrated, for the first time, an approach to detect autoimmune antibodies in patients with autoimmune disorders via P-ELISA. Bullous pemphigoid (BP) is a kind of autoimmune disease, the major epitope of type XVII collagen (BP180 or BPAG2) of a dermoepidermal junction called non-collagenous 16A domain (NC16A), which is identified as a marker that is recognized by autoantibodies in patients. In this P-ELISA system, only 2 μL of serum or blister fluid and 70 min were required to detect anti-NC16A autoimmune antibodies, demonstrating that this model may be applicable to many other autoimmune diseases, such as lupus erythematosus, or scleroderma [101]. **d** Mu et al. [110] developed a multiplex microfluidic paper-based immunoassay for the diagnosis of hepatitis C virus infection. The Center for Disease Control (CDC) has indicated that it will accept a first-line diagnosis via the serologic detection of IgG antibody against HCV (anti-HCV) by ELISA. A second assay using a recombinant immunoblot assay (RIBA) approach is imperative to circumvent the false-positive bias of ELISA and confirm the diagnostic result. What is most striking from this study was the fact that the required serum volume for testing was as low as 6 nL per detection zone, and in practice, only 0.3 μL of 50-fold diluted serum was needed. This is approximately 2000 times less than conventional ELISA and RIBA approaches, i.e., 10 and 20 μL serum volumes, respectively. A craft punch is an inexpensive tool (less than $2), and an array of them can be assembled for parallel fabrication. The total assay time to analyze patient serum was just 30 min. This is half of the time required for standard ELISA and one-twelfth of the time required for RIBA. P-ELISA is also useful in infectious disease detection [110]. **e** Wang et al. [26] demonstrated P-ELISA for the diagnosis of dengue virus infection in both buffer system and human serum, and it recognized different serotypes by different antigens of the same serotype. Paper-based indirect ELISA was developed specific to dengue virus serotype-2 by using a therapeutically based monoclonal antibody with excellent sensitivity and specificity and short operation duration (<1 h) [26]. **f** The detection range for bacteria is between 500 CFU/mL and 5000 CFU/mL. The advantage of the immunodisk sensor is that it does not require any preprocessing of biological sample and is capable of whole-cell bacterial detection. (1) 5×10^3 CFU/mL, (2) 4×10^3 CFU/mL, (3) 3×10^3 CFU/mL, (4) 0.5×10^3 CFU/mL of *S. aureus*, (5) buffer (control with no pathogens), and (6) 5×10^3 CFU/mL of *S. aureus* (control with AuNPs treated in the same manner as conjugation but with no antibodies) [111]

(Table 3.1) [100]. Murdock et al., interestingly enough, reported a human emotion stress marker, neuropeptide Y (NPY), that could be monitored with a paper-based tool. Using anti-NPY IgG, minute test zones as small as 3 mm in diameter, and solution volumes as low as 1.5 μL, this paper-based bioanalysis showed great promise for use in the field and demonstrated particular relevance for the diagnosis and treatment of post-traumatic stress disorder (PTSD), a feature of significant interest regarding personnel returning from military deployment (Table 3.1) [106]. Apilux et al. developed a novel automated μPAD to detect human chorionic gonadotropin (hCG) in urine samples. This μPAD, designed on a piece of nitrocellulose membrane patterned with an inkjet printer, required a single sample application to perform sequential and automatic steps of an ELISA. Necessary reagents were spotted onto the microfluidic circuit at proper locations for sequential, multistep ELISA. The authors employed a digital camera to record reactive color intensity, thus enabling them to monitor hCG presence in urine (Table 3.1) [107]. In summary, paper-based ELISA has been successfully used in lieu of conventional ELISA and offers a highly suitable replacement in a growing list of arenas.

3.2.5.3 Paper-Based Pathogen Diagnostics

Paper has also been successfully used as a platform for the detection of pathogens ranging from hepatitis B virus (HBV) in clinical serum samples to food-borne pathogens like *Listeria monocytogenes.* Dineva et al. used (RT)-PCR to amplify the specific molecule message of HBV from patients' plasma while taking advantage of dipstick assays through the sandwich hybridization to indicate the virus level, observed via naked eyes (Table 3.1) [108]. Hepatitis B surface antigen (HBsAg) was detectable in serum by using P-ELISA in a three-dimensional (3D) μPAD that required only a 2 μL sample and that stored all necessary reagents in the device itself [109]. HCV detection in a microfluidic paper-based immunoassay was achieved by Mu et al., who developed a multiplex assay to replace traditionally used first-line diagnostics. By benefit of the multiple efficiencies inherent in paper-based devices, Mu's tool is capable of benefitting large numbers of unwitting carriers (Fig. 3.8d) [110]. Rohrman et al. used layers of both paper and plastic to develop a lateral flow test device that could provide an inexpensive, rapid, and easy-to-use POC device to perform recombinant polymerase amplification of human HIV DNA in resource-limited areas. To detect dengue virus, Lo et al. applied loop-mediated isothermal amplification of dengue virus serotype-2 RNA followed by fluorescently based detection at the molecular level using paper [19]. Wang et al. also demonstrated successful diagnosis of dengue fever (serotype-2) using either P-ELISA (indirect) or lateral flow immunoassays (Fig. 3.8e) [26]. Wang and colleagues detected dengue virus in human serum from infected patients via capture ELISA of immunoglobulin M (IgM) antibody. IgM is a specific marker related to primary dengue infection. Li et al. described a multiplexed and disk-shaped μPAD capable of detecting *Pseudomonas aeruginosa* and *Staphylococcus aureus* with immobilized antibody-conjugated AuNPs for colorimetric detection of both bacteria (Fig. 3.8f) [111]. Jokerst et al. described a μPAD capable of detecting several food-borne pathogenic bacteria (*Escherichia coli O157:H7*, *Salmonella typhimurium*, and *L. monocytogenes*) in water using a colorimetric assay that detected enzymes characteristically produced by these bacteria and their respective chromogenic substrates [112].

3.2.5.4 DNA-Based Assays in Paper

Paper has also been used for DNA-based assays. Lo et al. employed real-time accelerated reverse transcription loop-mediated isothermal amplification (RT-LAMP) to amplify dengue virus serotype-2 RNA and diagnose its presence

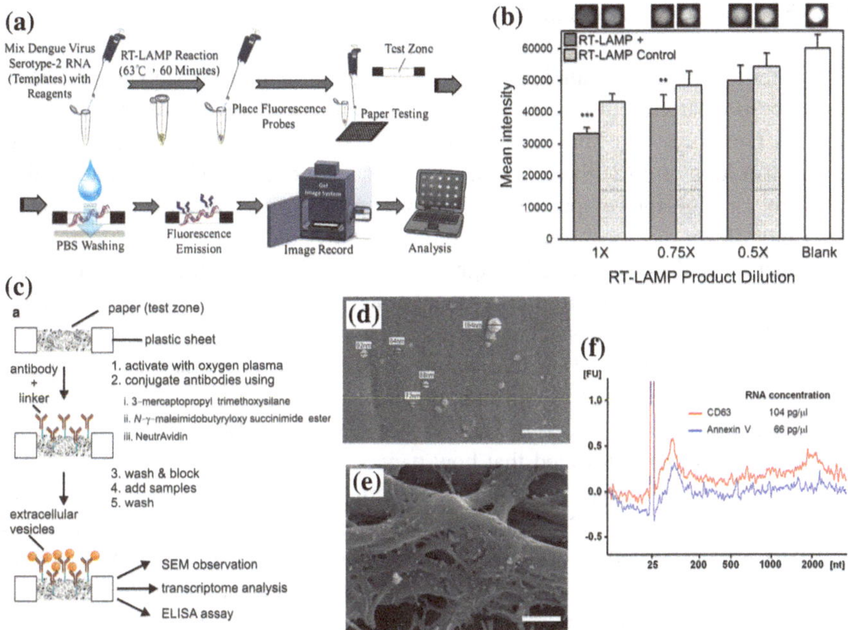

Fig. 3.9 Nucleotide-based assays in paper. **a** Schematic of the entire procedure for performing the diagnosis of dengue virus serotype-2 RNA in paper. **b** Dengue virus serotype-2 RNA (with a virus particle concentration of approximately 6000 PFU/mL) was transcribed reversely and then amplified via RT-LAMP for one hour at 63 °C. The amplified product was stained with U-safe fluorescence probes and could be used to detect with a dilution fold of 0.75 ($N = 8$; mean ± standard deviation) in paper-based test zones. *p, 0.05; **p, 0.01; ***p, 0.001 (Student's t-test), indicating statistically significant differences compared with the control group. **a–b** Were adapted from Lo et al. [19]. **c** A schematic showing the development of paper-based assays for isolation and characterization of extracellular vehicles (EVs). The paper surface was first activated via treatment with oxygen plasma and reacted with the following molecules: 3-mercaptopropyl trimethoxysilane, N-c-maleimidobutyryloxy succinimide ester, and NeutrAvidin. Biotinyl vesicle-specific antibodies (e.g., anti-CD63 antibody or anti-Annexin V antibody) were added and conjugated to previously modified paper. EVs from serum sample **d** and aqueous humor samples **e** were captured by anti-CD63-coated paper and examined via scanning electron microscope (SEM) images. **f** Total RNA was extracted from serum EVs captured on anti-CD63 antibody- or Annexin V-coated filter papers and analyzed with a bioanalyzer (Agilent Technologies, Santa Clara, CA). The data show the intensities (y-axis, in arbitrary units) and size (x-axis, in nucleotides (nt)) distributions of fluorescently labeled RNA. **c–f** were adapted from Chen et al. [113]

using a paper platform. This underscored the great potential for paper diagnostics by demonstrating an inexpensive DNA approach to detect dengue fever without a polymerase chain reaction (PCR) step (Fig. 3.9a and 3.9b) [19]. Chen et al. developed an immunoaffinity device using paper that could isolate and characterize extracellular vesicles (Fig. 3.9c) and, thus, successfully isolate them from 72 μL serum and aqueous humor samples (Fig. 3.9d, e). Moreover, Chen et al. demonstrated the ability to isolate RNA from extracellular vesicles captured in such a paper-based immune affinity device (Fig. 3.9f) [113], thus further displaying the capacity of paper platforms for DNA-based assays.

3.2.5.5 Cell-Based Assays in Paper

The potential of every candidate drug requires an evaluative step in cell-based assays before entering in vivo studies. Existing cell-based assays are primarily deficient in two ways: (1) The translational gaps between in vitro and in vivo studies are too long and (2) the efficiency and cost is too great. The current standard platform for high-throughput studies, 96-well plates, only allows cells to grow as a 2D monolayer, a process that poorly mimics physiological conditions including 3D structure and nutrient and biosignal gradients. Paper, on the other hand, provides a defined 3D structure and costs significantly less than conventional microplates. By stacking and destacking layers of paper with cells, the 3D nature of the substrate allowed Derda et al. to examine 3D cell cultures under varying oxygen and nutrient gradients and to more appropriately analyze cell growth and molecular responses, including those to HIF, VEGF, and more (Table 3.2) [21]. Further, Derda et al. also demonstrated that how papers wax-patterned with a 96-well format could provide a high-throughput platform for monitoring cell migration and cell growth (Table 3.2) [22]. Deiss et al. also reported using a unique holder that affixes a wax-patterned 96-well format on papers in order to study the cytotoxic effects of soluble molecules (Fig. 3.10a), including phenylarsine oxide (PAO) and cyclophosphamide (CPA), on human breast cancer cells in a 3D cell culture environment that more closely resembles the physiological milieu (Fig. 3.10b) [18]. Using Teflon-patterned paper, Deiss et al. were able to perform flow-through synthesis of peptides (Fig. 3.10c) and created a paper-based peptide array to monitor the different adhesion abilities of cells to different peptides in a high-throughput manner (Fig. 3.10d and 3.10e) [114]. These studies demonstrate that paper is a suitable high-throughput platform for monitoring the responses of cells to treatment with a variety of stimuli, be they peptides, small molecules, or glycans, which fortifies the argument for using paper to ameliorate the economic woes of the pharmaceutical industry. Clearly, high-throughput testing can be carried out on paper-based 3D cell cultures for the successful study of drug-related response.

Not only can paper-based platforms be used for eukaryotic cell culture, but also Funes-Huacca et al. demonstrated that they are suitable for prokaryotic cell culture study as well [115]. They showed that a portable device designed for the growth of bacteria or amplification of bacteriophages could be created using simple materials, including tape, sheets of paper patterned by hydrophobic printer ink, and a polydimethylsiloxane (PDMS) membrane. Using this device, Funes-Huacca and colleagues demonstrated comparable growth of *E. coli* and the amplification of the phage M13 in this device to that measured on agar plates and in shaking cultures. As a detection/diagnostic tool, this device was capable of detecting a low number of *E. coli* (1–10 CFU in 100 mL), and using reporter-gene-carrying *E. coli*, the device could be used to detect the concentration of arabinose (Table 3.2). Further, a similar paper-based device (Fig. 3.10f) has been used to successfully perform the Kirby–Bauer disk diffusion test to show that the antibiotic susceptibility of several strains of *E. coli* and *S. typhimurium,* as measured by staining zones of inhibited growth, is comparable to results found when using traditional agar dish cultures (Fig. 3.10g) [117].

Table 3.2 Current advances in cell-based assays on paper

Reference figures	Key breakthrough	Result demonstration and application	High throughput		
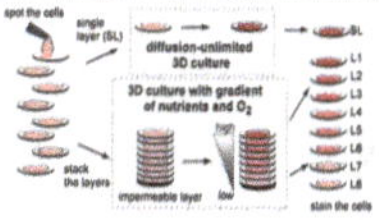	Use of paper for 3D cell culture and stacking multilayers of paper with cells to create tissue-based cell culture	Monitor cell growth under nutrient gradient. Monitor HIF expression of cells under nutrient and oxygen gradient	X	X	[21]
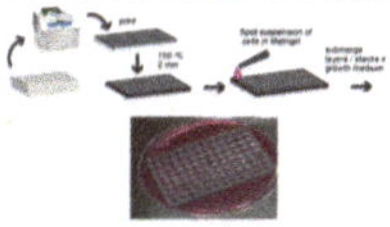	Use of wax-patterned paper to create 96-well format and stacking multilayers of paper for high-throughput assays in 3D cell culture	Monitor cell migration under nutrient and oxygen gradient Diffusion of oxygen and nutrients occurs through the wax-patterned paper	O	X	[22]
	Use of custom-made 96-ell holder to isolate the cell-containing zones from each other on wax-patterned paper	Monitor cytotoxicity of water-soluble compounds in 3D cell culture and show different results compared with 2D cell culture	O	X	[18]
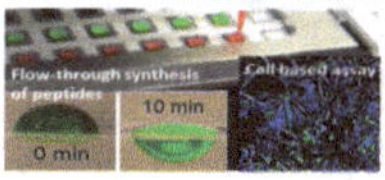	Direct synthesis of peptides on paper with 96-well format and production of peptide arrays for cell-based assays	Create Teflon-patterned paper by automation Perform flow-through synthesis of peptides by automation Monitor cell adhesion to different peptides on paper-based peptide array	O	O	[114]
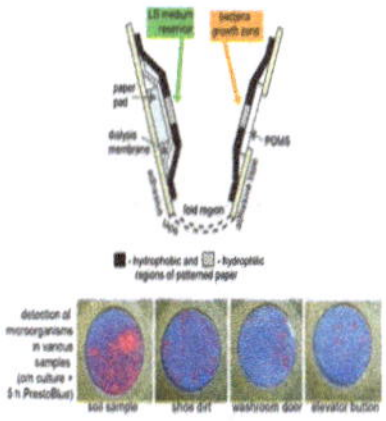	Use of paper and tape to fabricate a portable device for culture of phage and bacteria in replacing traditional Petri dish	Use paper and tape to fabricate a portable and self-contained device for culture of phage and bacteria. Note similar bacteria and phage growth in portable devices to the growth in standard culture. The concentration of bacteria can be estimated using a fluorescent or colorimetric readout, and image analysis could be performed by smartphone. Microorganisms from environmental samples could be detected using this device	X	X	[115]
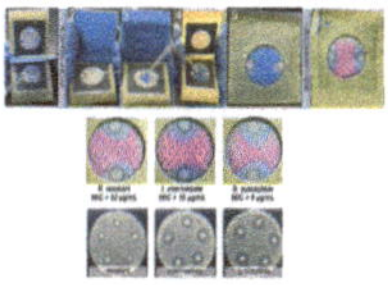	Use of paper- and tape-based portable culture device for antibiotic susceptibility test of bacteria	Show comparable bacterial antibody susceptibility to results attainable with traditional agar plates	X	X	[117]

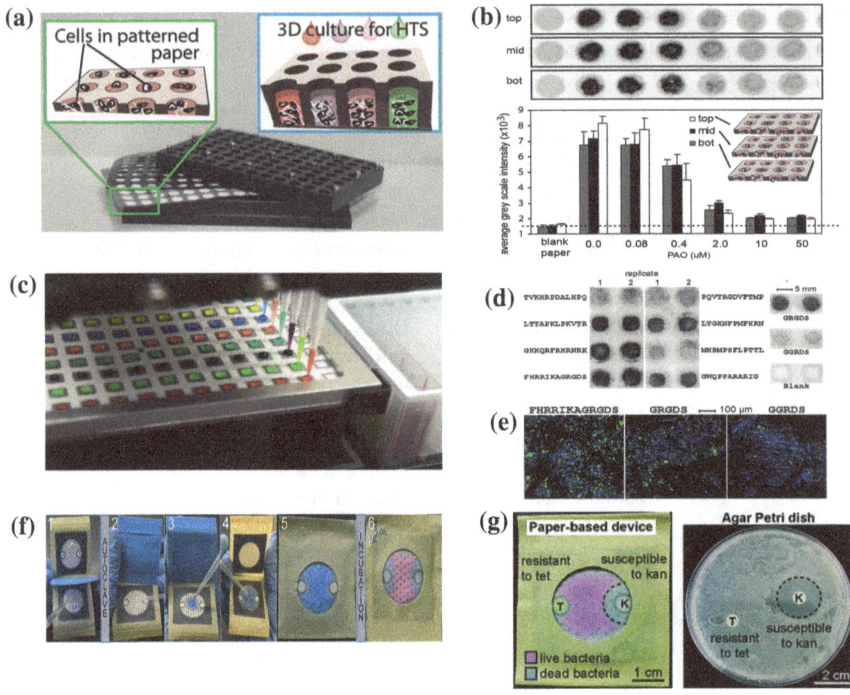

Fig. 3.10 Cell-based assays in paper. **a** Stacking wax-patterned papers designed with hydrophilic zones separated by hydrophobic barriers in a 96-well holder can prevent wicking of liquids between different wells through the connection of the paper and the 96-hole insert and allows for high-throughput testing to examine the effect of soluble compounds on paper-based 3D cell cultures. Adapted from Deiss et al. [18]. **b** MDA-MB-231 cells were loaded in paper with matrigel and exposed to different amounts of phenylarsine oxide (PAO) for a single day. Fluorescence imaging and analysis after calcein staining of 3D cultures with a fluorescence gel scanner were performed to monitor cytotoxicity of PAO on 3D cell culture. Adapted from Deiss et al. [18]. **c** Using automated spotting of different amino acids for peptide synthesis on paper and producing peptide array. Adapted from Deiss et al. [114]. **d** Imaging with a fluorescence gel scanner locates GFP fluorescence (dark areas) to monitor the adhesion of MDA-MB-231-GFP cells to known bioactive peptides synthesized on paper. **e** Confirmation of the results in **f** by confocal microscopy to monitor binding to GRGDS (positive control) and no binding to GGRDS (negative control). Adapted from Deiss et al. [114]. **g** A portable paper-based culture device for study of antimicrobial susceptibility. (1) Add LB media and close and sterilize by autoclave; (2) load the antibiotics in each A-zone and seal the device for storage; (3) load PrestoBlue™ to the C-zone and remove the blue protective band; (4) add *E. coli* K12 ER2738 to the C-zone; (5) close the device and incubate at 37°C for 18 h; (6) read the results. Adapted from Deiss et al. [117]. **h** Antibiotic susceptibility test was performed with different strains of *E. coli* K12 ER2738 that were resistant, intermediate, or susceptible to ampicillin on portable paper-based culture device and shows comparable results with that performed on traditional agar plates. Adapted from Deiss et al. [117]

3.3 Thread-/Cotton-Based Microfluidics

In addition to the disposable platforms mentioned earlier, thread has also appeared as an interesting alternative substrate for use in microfluidic applications. Like paper-based devices, thread is porous, inexpensive, and universally available. Thread's surface characteristics promote fluidic transport via capillary action, i.e., without the need of external forces or pumps [7]. A few research groups have demonstrated the capacity for threads to be used in immunochromatographic assays (ICAT) [118], chemical synthesis and sensing [119], and electrophoretic separations coupled to electrochemical detection [120].

Zhou et al. [118] reported the development of ICAT using cotton threads and nylon fiber bundles in place of pads and membranes. Their ICAT comprised a sandwich assay performed on a cotton thread knotted to a nylon fiber bundle, both of which were previously coated with recognition antibodies against a specific target analyte (Figs. 3.11a and 3.11b). Using this technique, visible results presented in a matter of minutes and could be quantified with a flatbed scanner. Cotton threads in this study (essentially twisted cellulose fibers) were treated with plasma to make them hydrophilic so that spontaneous wicking of aqueous solutions could

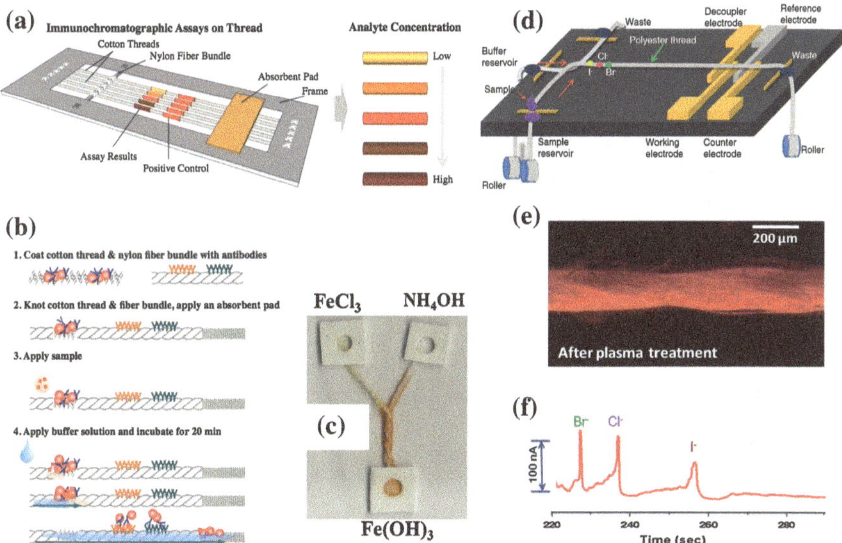

Fig. 3.11 Examples of thread-based devices and microfluidic applications: **a** device for immunochromatographic assays; **b** operation scheme of a bioassay on a thread; **c** device for chemical synthesis; **d** device for electrophoresis coupled with amperometric detection; **e** example of a polyester thread after plasma treatment; and **f** typical electropherogram recorded on a thread. Images were reprinted with permission from Zhou et al. [118], Banerjee et al. [119], and Wei et al. [120]

occur. Using lateral flow, the authors demonstrated multiplex measurements of C-reactive protein, leptin, and osteopontin with LOD values in the pM range.

Banerjee et al. [119] showed that chemical syntheses and sensing could be implemented using threads as microchannels by demonstrating the synthesis of brown ferric hydroxide in a Y-geometry thread reactor (see Fig. 3.11c). Passively mixing ferric chloride and ammonium hydroxide, the authors demonstrated synthesis efficiency on threads of approximately 84 %. Moreover, in addition to synthesis, they successfully performed colorimetric assays for BSA and glucose in blood plasma.

In an experiment that creatively coupled electrophoretic separation with amperometric detection, Wei et al. [120] proposed that the design of their thread-based electrophoresis device was similar to the structure of a traditional stringed musical instrument called a dulcimer. The "strings" were used as the sample routes, and the protruding nuts were used as the electrical contacts (see Fig. 3.11d). Notably, the authors showed that oxygen plasma treatment improved wettability and surface quality (Fig. 3.11e) and the measured electrical currents on plasma-treated threads were 10 times greater than the currents on native threads. Amperometric detection in these experiments was performed using a conventional three-electrode cell integrated with a gold decoupler. They

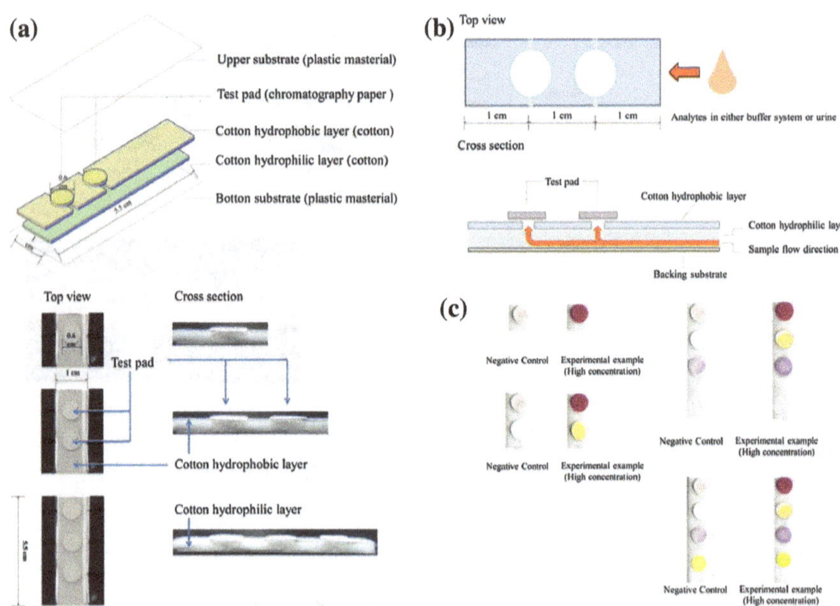

Fig. 3.12 Cotton-based diagnostic device. **a** Schematic diagram of lateral flow cotton-based diagnostic device. **b** Top view and cross-sectional diagrams indicated the sample flow direction. **c** The device has been shown for the diagnosis of BSA, nitrite, and uric acid assays in a buffer system [121]

achieved separations of anions (Br^-, Cl^-, and I^-) and catecholamines (catechol and dopamine) within 5 min, with LOD estimated at approximately 0.3 mM (see Fig. 3.11f).

Although cotton-based threads have inherent capillary action, their small surface areas have the fact of uneasy readouts for observing colorimetric detections and their original mechanical property is not favorable for users to operate due to too soft to dip fluently, if they do not package specially. Recently, Lin et al. [121] demonstrated a combination design that employed cotton as flow channel and chromatography paper as reaction zone. Cotton-based diagnostic devices are clearly advantageous due to ease of use, inexpensiveness, accuracy, and the fact that they are a US FDA-approved material. While still under development, current cotton instruments have the potential to perform additional diagnostic functions not yet explored (see Fig. 3.12).

3.4 Commercialization of Low-cost Microfluidic Devices for Clinical Diagnostics

Among the technologies described in this review, several demonstrate great potential for commercial production. They are inexpensive, rapid, portable, and reliable tools that demonstrate acutely particular suitability in resource-poor and remote settings. In a critical review published in 2012, Chin and colleagues [4] detail the primary companies responsible for the existing arena of commercialized POC diagnostic devices.

Because of its obvious advantages, paper, as a microfluidic platform, has received the lion's share of attention regarding commercialization. Governmental and nongovernmental organizations, such as the Sentinel Bioactive Paper Network, the Bioresource Processing Research Institute of Australia (BioPRIA), the Program for Appropriate Technology in Health (PATH), Diagnostic for All (DFA), and the Bill & Melinda Gates Foundation, have all played an important role in the development and commercialization of μPADs for diagnostic applications in resource-poor settings. Already, the μPAD described in this review for liver function testing is in an advanced stage of development for mass production by the nonprofit enterprise DFA, and a μPAD for blood typing is also under consideration for commercialization by BioPRIA.

For a variety of reasons, including well-established fabrication technologies such as injection molding and hot embossing, the use of plastics for the production of microfluidic devices is, today, more mature than the use of paper. Plastics allow for easy integration of different fluidic components for sample pretreatment, volume control, sample mixing, and signal detection. In short, plastic platforms provide a sample-in/answer-out capability that is desirable for POC diagnostics. Different companies have commercialized polymeric devices for clinical diagnostics, where the sample including whole blood, tears, and urine can be handled under centrifugal, capillary, mechanical, or gravitational forces. For the products available in the market, the signal is processed by absorbance, colorimetric, electrochemical, or fluorescence detection. The pioneering companies in this area include Abaxis (http://www.abaxis.com/), Alere (http://alere-technologies.com/), Focus Diagnostics (http://www.focusdx.com/), Micronics (https://www.micronics.net/), Mbio Diagnostics Inc. (http://mbiodx.com/), TearLab (http://www.tearlab.com/), and Zyomyx (http://www.zyomyx.com/) [4]. In addition to μPADs and polymeric devices, toner-based platforms have also demonstrated great potential for commercialization especially where the availability of funds is quite limited [5].

Despite all the recent advances and improvements in the production of low-cost microfluidic devices, commercialization remains underexploited. Finding a more ready path from proof of concept to the marketplace can be improved with greater cooperation between research and industry. Necessary requirements for marketability include suitability for mass production as well as demonstrated feasibility and reliability, and approval for clinical use understandably relies on significant testing and regulatory approval, but life-changing diagnostics awaits in the wings, and industry must be persuaded to ratchet up the process of turning academic prototypes into real products.

3.5 Concluding Remarks

In the period covered by this review, a remarkable and increasing trend in the use of low-cost substrates to develop microfluidic devices dedicated to diagnostic and clinical assays has been observed. Paper substrates have dominated this field, largely because of their capacity to be used as POC diagnostic devices for resource-poor settings. Nevertheless, contributions from plastic platforms should not be disregarded. The development of diagnostic microfluidic devices produced on affordable platforms has been enhanced in many ways by new and enhanced detection approaches, integration and automation of analytical steps, inclusion of functionalities (self-power and easy readout, for example), and applications suitable for real-life scenarios. Certainly, many of the microfluidic devices described in this review have the potential to improve life quality by providing inexpensive, rapid, portable, and reliable diagnostics. The remaining bottleneck in this process is translationally streamlining productive and value-laden research to real-life deployment and the marketplace.

References

1. Yager P, Edwards T, Fu E, Helton K, Nelson K, Tam MR, Weigl BH (2006) Nature 442:412–418
2. Foudeh AM, Didar TF, Veres T, Tabrizian M (2012) Lab Chip 12:3249–3266
3. Martinez AW, Phillips ST, Whitesides GM, Carrilho E (2010) Anal Chem 82:3–10
4. Chin CD, Linder V, Sia SK (2012) Lab Chip 12:2118–2134
5. Coltro WKT, de Jesus DP, da Silva JAF, do Lago CL, Carrilho E (2010) Electrophoresis 31:2487–2498
6. Yetisen AK, Akram MS, Lowe CR (2013) Lab Chip 13:2210–2251

7. Ballerini DR, Li X, Shen W (2012) Microfluid Nanofluid 13:769–787
8. Nery EW, Kubota LT (2013) Anal Bioanal Chem 405:7573–7595
9. Martinez AW, Phillips ST, Butte MJ, Whitesides GM (2007) Angew Chem Int Ed 46:1318–1320
10. Hu J, Wang SQ, Wang L, Li F, Pingguan-Murphy B, Lu TJ, Xu F (2014) Biosens Bioelectron 54:585–597
11. Carrilho E, Phillips ST, Vella SJ, Martinez AW, Whitesides GM (2009) Anal Chem 81:5990–5998
12. Martinez AW, Phillips ST, Whitesides GM (2008) Proc Natl Acad Sci USA 105:19606–19611
13. Lutz B, Liang T, Fu E, Ramachandran S, Kauffman P, Yager P (2013) Lab Chip 13:2840–2847
14. Martinez AW, Phillips ST, Nie ZH, Cheng CM, Carrilho E, Wiley BJ, Whitesides GM (2010) Lab Chip 10:2499–2504
15. Ellerbee AK, Phillips ST, Siegel AC, Mirica KA, Martinez AW, Striehl P, Jain N, Prentiss M, Whitesides GM (2009) Anal Chem 81:8447–8452
16. Vella SJ, Beattie P, Cademartiri R, Laromaine A, Martinez AW, Phillips ST, Mirica KA, Whitesides GM (2012) Anal Chem 84:2883–2891
17. Pollock NR, Rolland JP, Kumar S, Beattie PD, Jain S, Noubary F, Wong VL, Pohlmann RA, Ryan US, Whitesides GM (2012) Sci Transl Med 4(152r):a129
18. Deiss F, Mazzeo A, Hong E, Ingber DE, Derda R, Whitesides GM (2013) Anal Chem 85:8085–8094
19. Lo SJ, Yang SC, Yao DJ, Chen JH, Tu WC, Cheng CM (2013) Lab Chip 13:2686–2692
20. Cheng CM, Martinez AW, Gong JL, Mace CR, Phillips ST, Carrilho E, Mirica KA, Whitesides GM (2010) Angew Chem Int Ed 49:4771–4774
21. Derda R, Laromaine A, Mammoto A, Tang SKY, Mammoto T, Ingber DE, Whitesides GM (2009) Proc Natl Acad Sci USA 106:18457–18462
22. Derda R, Tang SKY, Laromaine A, Mosadegh B, Hong E, Mwangi M, Mammoto A, Ingber DE, Whitesides GM (2011) PLoS ONE 6:e18940
23. Tsai TT, Shen SW, Cheng CM, Chen CF (2013) Sci Technol Adv Mater 14:044404
24. Cheng CM, Mazzeo AD, Gong JL, Martinez AW, Phillips ST, Jain N, Whitesides GM (2010) Lab Chip 10:3201–3205
25. Hsu MY, Yang CY, Hsu WH, Lin KH, Wang CY, Shen YC, Chen YC, Chau SF, Tsai HY, Cheng CM (2014) Biomaterials 35:3729–3735
26. Wang HK, Tsai CH, Chen KH, Tang CT, Leou JS, Li PC, Tang YL, Hsieh HJ, Wu HC, Cheng CM (2014) Adv Healthc Mater 3:187–196
27. Carrilho E, Martinez AW, Whitesides GM (2009) Anal Chem 81:7091–7095
28. Songjaroen T, Dungchai W, Chailapakul O, Laiwattanapaisal W (2011) Talanta 85:2587–2593
29. Dungchai W, Chailapakul O, Henry CS (2011) Analyst 136:77–82
30. Zhang AL, Zha Y (2012) Aip Adv 2:022171
31. Zhong ZW, Wang ZP, Huang GXD (2012) Microsyst Technol 18:649–659
32. Nie JF, Zhang Y, Lin LW, Zhou CB, Li SH, Zhang LM, Li JP (2012) Anal Chem 84:6331–6335
33. Ge L, Wang SM, Song XR, Ge SG, Yu JH (2012) Lab Chip 12:3150–3158
34. Liu H, Crooks RM (2011) J Am Chem Soc 133:17564–17566
35. Tian JF, Jarujamrus P, Li LZ, Li MS, Shen W (2012) ACS Appl Mater Interf 4:6573–6578
36. Lewis GG, DiTucci MJ, Baker MS, Phillips ST (2012) Lab Chip 12:2630–2633
37. He QH, Ma CC, Hu XQ, Chen HW (2013) Anal Chem 85:1327–1331
38. Glavan AC, Martinez RV, Maxwell EJ, Subramaniam AB, Nunes RMD, Soh S, Whitesides GM (2013) Lab Chip 13:2922–2930
39. Schilling KM, Jauregui D, Martinez AW (2013) Lab Chip 13:628–631
40. Nie JF, Liang YZ, Zhang Y, Le SW, Li DN, Zhang SB (2013) Analyst 138:671–676
41. Chitnis G, Ding ZW, Chang CL, Savran CA, Ziaie B (2011) Lab Chip 11:1161–1165
42. Ornatska M, Sharpe E, Andreescu D, Andreescu S (2011) Anal Chem 83:4273–4280
43. Peng P, Summers L, Rodriguez A, Garnier G (2011) Colloids Surf B-Biointerf 88:271–278

44. Tseng SC, Yu CC, Wan DH, Chen HL, Wang LA, Wu MC, Su WF, Han HC, Chen LC (2012) Anal Chem 84:5140–5145
45. Yuan JP, Gaponik N, Eychmuller A (2012) Anal Chem 84:5047–5052
46. Yu A, Shang J, Cheng F, Paik BA, Kaplan JM, Andrade RB, Ratner DM (2012) Langmuir 28:11265–11273
47. Liang J, Wang YY, Liu B (2012) RSC Adv 2:3878–3884
48. Yildiz UH, Alagappan P, Liedberg B (2013) Anal Chem 85:820–824
49. Noor MO, Shahmuradyan A, Krull UJ (2013) Anal Chem 85:1860–1867
50. Yu JH, Ge L, Huang JD, Wang SM, Ge SG (2011) Lab Chip 11:1286–1291
51. Wang SM, Ge L, Song XR, Yu JH, Ge SG, Huang JD, Zeng F (2012) Biosens Bioelectron 31:212–218
52. Wang SM, Ge L, Song XR, Yan M, Ge SG, Yu JH, Zeng F (2012) Analyst 137:3821–3827
53. Wang YH, Wang SM, Ge SG, Wang SW, Yan M, Zang DJ, Yu JH (2013) Anal Methods 5:1328–1336
54. Forster RJ, Bertoncello P, Keyes TE (2009) Ann Rev Anal Chem 2:359–385
55. Delaney JL, Hogan CF, Tian JF, Shen W (2011) Anal Chem 83:1300–1306
56. Shi CG, Shan X, Pan ZQ, Xu JJ, Lu C, Bao N, Gu HY (2012) Anal Chem 84:3033–3038
57. Ge L, Yan JX, Song XR, Yan M, Ge SG, Yu JH (2012) Biomaterials 33:1024–1031
58. Yan JX, Ge L, Song XR, Yan M, Ge SG, Yu JH (2012) Chem A Eur J 18:4938–4945
59. Wang SW, Ge L, Zhang Y, Song XR, Li NQ, Ge SG, Yu JH (2012) Lab Chip 12:4489–4498
60. Li WP, Li M, Ge SG, Yan M, Huang JD, Yu JH (2013) Anal Chim Acta 767:66–74
61. Xu YH, Lou BH, Lv ZZ, Zhou ZX, Zhang LB, Wang EK (2013) Anal Chim Acta 763:20–27
62. Xu YH, Lv ZZ, Xia Y, Han YC, Lou BH, Wang EK (2013) Anal Bioanal Chem 405:3549–3558
63. Yan JX, Yan M, Ge L, Yu JH, Ge SG, Huang JD (2013) Chem Commun 49:1383–1385
64. Dungchai W, Chailapakul O, Henry CS (2009) Anal Chem 81:5821–5826
65. Carvalhal RF, Kfouri MS, Piazetta MHD, Gobbi AL, Kubota LT (2010) Anal Chem 82:1162–1165
66. Nie ZH, Nijhuis CA, Gong JL, Chen X, Kumachev A, Martinez AW, Narovlyansky M, Whitesides GM (2010) Lab Chip 10:477–483
67. Nie ZH, Deiss F, Liu XY, Akbulut O, Whitesides GM (2010) Lab Chip 10:3163–3169
68. Liu H, Crooks RM (2012) Anal Chem 84:2528–2532
69. Rattanarat P, Dungchai W, Siangproh W, Chailapakul O, Henry CS (2012) Anal Chim Acta 744:1–7
70. Godino N, Gorkin R, Bourke K, Ducree J (2012) Lab Chip 12:3281–3284
71. Santhiago M, Wydallis JB, Kubota LT, Henry CS (2013) Anal Chem 85:5233–5239
72. Shiroma LY, Santhiago M, Gobbi AL, Kubota LT (2012) Anal Chim Acta 725:44–50
73. Santhiago M, Kubota LT (2013) Sens Actuators B-Chem 177:224–230
74. Noiphung J, Songjaroen T, Dungchai W, Henry CS, Chailapakul O, Laiwattanapaisal W (2013) Anal Chim Acta 788:39–45
75. Lu JJ, Ge SG, Ge L, Yan M, Yu JH (2012) Electrochim Acta 80:334–341
76. Ge SG, Ge L, Yan M, Song XR, Yu JH, Huang JD (2012) Chem Commun 48:9397–9399
77. Wang PP, Ge L, Yan M, Song XR, Ge SG, Yu JH (2012) Biosens Bioelectron 32:238–243
78. Zang DJ, Ge L, Yan M, Song XR, Yu JH (2012) Chem Commun 48:4683–4685
79. Liu H, Xiang Y, Lu Y, Crooks RM (2012) Angew Chem Int Ed 51:6925–6928
80. Li WP, Li L, Li M, Yu JH, Ge SG, Yan M, Song XR (2013) Chem Commun 49:9540–9542
81. Yu WW, White IM (2011) Proc SPIE 7911:791105
82. Yu WW, White IM (2013) Analyst 138:1020–1025
83. Ngo YH, Li D, Simon GP, Garnier G (2012) Langmuir 28:8782–8790
84. Chen YY, Cheng HW, Tram K, Zhang SF, Zhao YH, Han LY, Chen ZP, Huan SY (2013) Analyst 138:2624–2631
85. Abbas A, Brimer A, Slocik JM, Tian LM, Naik RR, Singamaneni S (2013) Anal Chem 85:3977–3983

86. Lewis GG, DiTucci MJ, Phillips ST (2012) Angew Chem Int Ed 51:12707–12710
87. Tian LM, Morrissey JJ, Kattumenu R, Gandra N, Kharasch ED, Singamaneni S (2012) Anal Chem 84:9928–9934
88. Ge L, Wang PP, Ge SG, Li NQ, Yu JH, Yan M, Huang JD (2013) Anal Chem 85:3961–3970
89. Wang PP, Ge L, Ge SG, Yu JH, Yan M, Huang JD (2013) Chem Commun 49:3294–3296
90. Fu E, Liang T, Houghtaling J, Ramachandran S, Ramsey SA, Lutz B, Yager P (2011) Anal Chem 83:7941–7946
91. Hwang H, Kim SH, Kim TH, Park JK, Cho YK (2011) Lab Chip 11:3404–3406
92. Lutz BR, Trinh P, Ball C, Fu E, Yager P (2011) Lab Chip 11:4274–4278
93. Fu EL, Ramsey S, Kauffman P, Lutz B, Yager P (2011) Microfluid Nanofluid 10:29–35
94. Schilling KM, Lepore AL, Kurian JA, Martinez AW (2012) Anal Chem 84:1579–1585
95. Kwong P, Gupta M (2012) Anal Chem 84:10129–10135
96. Jahanshahi-Anbuhi S, Chavan P, Sicard C, Leung V, Hossain SMZ, Pelton R, Brennan JD, Filipe CDM (2012) Lab Chip 12:5079–5085
97. Rezk AR, Qi A, Friend JR, Li WH, Yeo LY (2012) Lab Chip 12:773–779
98. Thom NK, Yeung K, Pillion MB, Phillips ST (2012) Lab Chip 12:1768–1770
99. Thom NK, Lewis GG, DiTucci MJ, Phillips ST (2013) RSC Adv 3:6888–6895
100. Yamada K, Takaki S, Komuro N, Suzuki K, Citterio D (2014) Analyst 139:1637–1643
101. Hsu CK, Huang HY, Chen WR, Nishie W, Ujiie H, Natsuga K, Fan ST, Wang HK, Lee JYY, Tsai WL, Shimizu H, Cheng CM (2014) Anal Chem 86:4605–4610
102. Martinez AW (2011) Bioanalysis 3:2589–2592
103. Klasner SA, Price AK, Hoeman KW, Wilson RS, Bell KJ, Culbertson CT (2010) Anal Bioanal Chem 397:1821–1829
104. Wong AP, Gupta M, Shevkoplyas SS, Whitesides GM (2008) Lab Chip 8:2032–2037
105. Matsuura K, Chen KH, Tsai CH, Li WQ, Asano Y, Naruse K, Cheng CM (2014) Microfluid Nanofluid 16:857–867
106. Murdock RC, Shen L, Griffin DK, Kelley-Loughnane N, Papautsky I, Hagen JA (2013) Anal Chem 85:11634–11642
107. Apilux A, Ukita Y, Chikae M, Chailapakul O, Takamura Y (2013) Lab Chip 13:126–135
108. Dineva MA, Candotti D, Fletcher-Brown F, Allain JP, Lee H (2005) J Clin Microbiol 43:4015–4021
109. Liu XY, Cheng CM, Martinez AW, Mirica KA, Li XJ, Phillips ST, Mascarenas M, Whitesides GM (2011) 2011 IEEE 24th international conference on micro electro mechanical systems (Mems), pp 75–78
110. Mu X, Zhang L, Chang SY, Cui W, Zheng Z (2014) Anal Chem 86:5338–5344
111. Li CZ, Vandenberg K, Prabhulkar S, Zhu XN, Schneper L, Methee K, Rosser CJ, Almeide E (2011) Biosens Bioelectron 26:4342–4348
112. Jokerst JC, Emory JM, Henry CS (2012) Analyst 137:24–34
113. Chen CC, Lin BR, Wang HK, Fan ST, Hsu MY, Cheng CM (2014) Microfluid Nanofluid 16:849–856
114. Deiss F, Matochko WL, Govindasamy N, Lin EY, Derda R (2014) Angew Chem Int Ed 53:6374–6377
115. Funes-Huacca M, Wu A, Szepesvari E, Rajendran P, Kwan-Wong N, Razgulin A, Shen Y, Kagira J, Campbell R, Derda R (2012) Lab Chip 12:4269–4278
116. Chen YH, Kuo ZK, Cheng CM (2015) Trends Biotechnol 33:4–9
117. Deiss F, Funes-Huacca ME, Bal J, Tjhung KF, Derda R (2014) Lab Chip 14:167–171
118. Zhou GN, Mao X, Juncker D (2012) Anal Chem 84:7736–7743
119. Banerjee SS, Roychowdhury A, Taneja N, Janrao R, Khandare J, Paul D (2013) Sens Actuator B Chem 186:439–445
120. Wei YC, Fu LM, Lin CH (2013) Microfluid Nanofluid 14:583–590
121. Lin SC, Hsu MY, Kuan CM, Wang HK, Chang CL, Tseng FG, Cheng CM (2014) Sci Rep 4:6976

Chapter 4
Glucose Sensor and Its Potential Directions

4.1 Overview

With the rise of the aging population, lifestyle changes, and heavy investments in infectious (acute) diseases, it has been difficult to increase focus and expenditure on the treatment of chronic diseases, such as heart disease, cancer, stroke, diabetes, and arthritis, all of which affect societies worldwide [1]. The World Health Organization (WHO) estimated that major chronic diseases account for 40 % of the global disease burden, and their contribution will rise to 60 % of the global disease burden by 2020 [2]. Patients with these chronic diseases are often afflicted with severe disease combinations or clusters of chronic conditions, and some diseases may result in significant, long-standing, and costly disabilities. Diabetes, for example, is often accompanied by kidney failure, non-traumatic lower extremity amputations, and blindness [3, 4]. In 2005, 21 % (approximately 63 million) of Americans had more than 1 chronic condition or had multiple illnesses or impairments expected to last a year or longer [3]. Unfortunately, current strategies/policies for controlling the prevalence of chronic diseases are not yet in place because there are no well-established healthcare systems or guidelines aimed at reducing the massive cost of chronic disease healthcare burdens from diagnosis, treatment, and care. Among these chronic diseases, cardiovascular disease, cancer, chronic obstructive pulmonary disease, and type-2 diabetes impose the most damage on our society [2]. The concept of diabetes management has, however, become more prevalent in recent years; in 2004, 97 % of private health plans focused on diabetes management [5]. In the USA, the healthcare expenditure for diabetes was approximately 245 billion dollars in 2012 and was approximately 2.3 times higher than all healthcare costs for people without diabetes [6]. In addition to physical care (prevention, diagnosis, and treatment), patients with diabetes are also afflicted with comorbid mental health disorders, such as major depressive disorder and generalized anxiety disorder [7, 8]. To ameliorate the prevalence of diabetes, the idea of

C.-M. Cheng et al., *In-Vitro Diagnostic Devices*, DOI 10.1007/978-3-319-19737-1_4

personal self-management is revolutionary; it helps afflicted individuals manage diabetes risk factors (e.g., obesity, genetic defects) and physical conditions, and facilitates prompt, appropriate medical care when glucose levels exceed the standard level. Accordingly, portable blood glucose meters, a classic representative for in vitro diagnostic devices, have become increasingly common as a home-based diagnostic tool because they provide flexibility and offer timely (<10 s for detection results) on-site detection of blood glucose levels. Three main components transduce blood glucose concentration into a readable signal in a glucose meter: (i) the biologic recognition elements that launch the redox reactions with glucose molecules in blood; (ii) a transducer that converts the redox reactions occurring in biologic recognition elements into a measurable signal; and (iii) a signal-processing platform that transduces the electrical signal into a readable result [9, 10]. To date, glucose oxidase-based and glucose dehydrogenase-based redox reactions are widely being applied in various brands of glucose meters [9] (Table 4.1). However, the glucose oxidase-based meter performance can be influenced by oxygen levels in the air. In 1996, the Americans with Disabilities Act (ADA) recommended that glucose meters comply with the suggestion for a <5 % maximum threshold of allowable bias (error) from reference methods for glucose concentrations among 1.6–22.2 mM [11]. In 2003, the International Organization for Standardization

Table 4.1 Commercially available glucose meters [9]

Manufacturer	Brand	Assay method	Minimal sample volume (uL)	Test time (second)	Assay range (mg/dL)	Hematocrit range (%)	Memory (results)
Abbott	FreeStyle freedom lite	GDH-PQQ	0.3	–5	20–500	15–65	400
AgaMatrix	WaveSense KeyNote	GOD	0.5	4	20–600	20–60	300
Arkray	Glucocard X-meter	GDH	0.3	5	10–600	30–52	360
Bayer	Ascensia contour	GDH-FAD	0.6	5	10–600	0–70	480
Bionime	Rightest GM300	GOD	1.4	8	20–600	30–55	300
Diabestic supply of suncoast	Advocate Redi-Code[a]	GOD	0.7	7	20–600	20–60	450
Diagnostic devices	Prodigy autocode	GOD	0.6	6	20–600	20–60	450
LifeScan	OneToucli UltraLink	GOD	1.0	5	20–600	30–55	500
Nova biomedical	Nova max	GOD	0.3	5	20–600	25–60	400
Roche	Accu-Chek aviva	GDH-PQQ	0.6	5	10–600	20–70	500

[a]This monitor has audio features to help visually impaired use

(ISO) suggested that 95 % of results should fall in the ±20 % range and ±15 mg/dL range for blood glucose >76 and ≤76 mg/dL, respectively [12]. However, glucose meters have two significant disadvantages: (i) they rely on a fingerstick to collect blood for each operation, which may induce infection and can be complicated for neonatal, elderly, and hemophobic individuals [13], and (ii) their diagnostic results are not suitable for continuous monitoring because machine designs do not provide the information and communications technology (ICT) necessary for recording and analysis. Interestingly, the technology of mobile phones provides promising options for continuous monitoring and telemedicine efforts. They are capable of acting as digital processing centers that can transduce detection signals from diagnostic/analytical devices into easily recognized results [14–18]. Further, monitoring results can be transmitted after processing to cloud-based servers, where physicians can review information and provide appropriate treatment guidelines regardless of geographic barriers [16, 19]. This is especially useful for optimizing diabetes management. Technological advances in wearable diagnostics have prompted the development of approaches that could significantly advance quality of life [20–23]. In regard to continuous glucose level monitoring, a noninvasive contact lens containing electrochemical measurement sensors is thought to be a most promising approach to managing monitoring needs. The contact lens-based glucose electrochemical measurement sensors provide a number of advantages far beyond conventional blood glucose meters: (i) they are based on detection via basal tear fluid rather than fingerstick blood samples, so the process is painless (a correlation between tear and blood glucose levels has been confirmed) [24]; (ii) the contact lens is a ubiquitous, FDA-approved medical device, and people are familiar with such products in their homes; (iii) contact lens fabrication is compatible with semiconductor fabrication techniques, so miniature electrical components and MEM devices can be mounted onto the surface of a contact lens; (iv) the barrier for mass production is comparatively low thanks to the adherence of matured semiconductor fabrications; and (v) most contact lens-based glucose electrochemical measurement sensors are compatible with ICT (i.e., mobile phone applications); hence, physicians can receive synchronous information when individuals operate the devices. Notably, the Sensimed AG company has commercialized a contact lens-based product (currently only approved in Europe)—the product SENSIMED Triggerfish. The Triggerfish is designed to monitor continuous ocular dimensional changes over 24 h for glaucoma management by using wireless pressure transducers mounted onto the surface of a contact lens [25]. In 2004, March et al. [26] discussed results regarding the primary trial of a noninvasive contact lens glucose sensor—they used a fluorescence probe to indicate the concentration of glucose. Chu et al. [27] developed a soft tear glucose biosensor that could be placed on the surface of a contact lens composed of polydimethylsiloxane (PDMS) and implanted into a rabbit eye. The measurement, however, had to rely on a benchtop potentiostat and a hard-wired connection to link the biosensor. Pandey et al. [28] proposed a wirelessly powered active contact lens for tear fluid measurement that did not obstruct wearer vision (Fig. 4.1). In this device, a loop antenna, power harvesting IC, and micro-LED were incorporated onto the

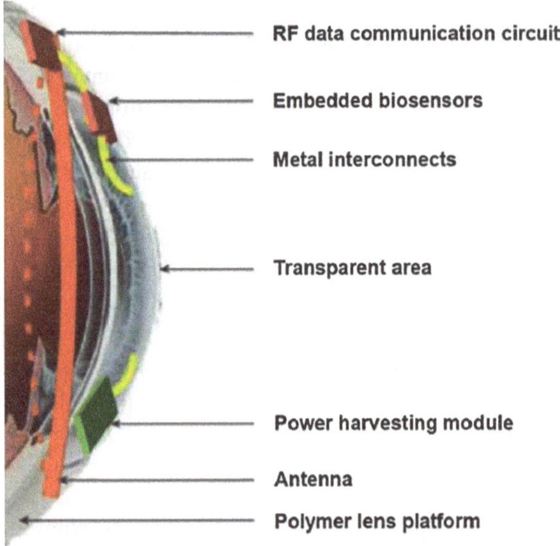

Fig. 4.1 A conceptual design for an active contact lens [28]

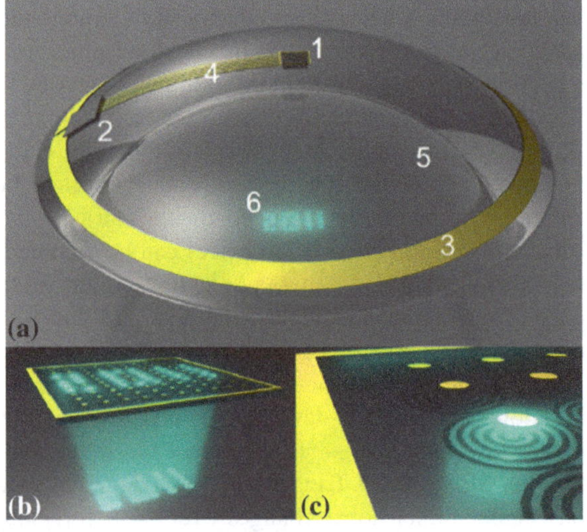

Fig. 4.2 Conceptual rendition of a multipixel contact lens display. **a** Image of a contact lens wih light emiting diode (LED) chip. **b** LED chip with 100 pixels. **c** Magnified view with one pixel activated, showing Fresnel lenses opposie each LED pixel [29]

surface of contact lens, proving that a lens powered wirelessly by radio frequency was feasible. The authors accomplished a single-pixel wireless design and tested its biocompatibility on a rabbit eye, while examining the relationship between antenna designs and operating distances under various physical conditions (Fig. 4.2) [29]. The operation distance for activating the contact lens LED on a rabbit eye was about 10 cm with a "horn" antenna and +35 dBm input power at 1.05 GHz. When saline film was used to cover the contact lens, the frequency declined significantly (from approximately 1.05 to approximately 1.8 GHz) and the operational range became shorter. The group further incorporated sophisticated

CMOS glucose sensors into a wirelessly powered active contact lens by integrating a sensor system (a loop antenna, wireless sensor interface chip, and a glucose sensor) and IC components (power management, readout circuitry, wireless communication interface), an LED driver, and energy storage capacitors in a CMOS chip) (Fig. 4.3) [30]. This contact lens performed a linear gain of 400 Hz/mM for glucose levels ranging between 0.05 and 1 mM (consumed 3 μW from a regulated on-chip 1.2-V supply). Tears are a complex extracellular fluid containing proteins/peptides, electrolytes, lipids, and metabolites from lacrimal glands, ocular surface epithelial cells, meibomian glands, goblet cells, and blood. They contain only low glucose concentrations (normal tear glucose levels range from 0.1 to 0.6 mM) [31, 32]. The authors above also developed a second-generation contact lens that contained dual glucose sensors and a telecommunication circuit (Fig. 4.4) [33]. The primary sensor was treated with active glucose oxidase (GOD), while the control sensor was treated with deactivated GOD. The company, Google, recently announced a report about devoting its resources to creating a smart contact lens impregnated with Novartis to assist with diabetes management [34]. This announcement aroused great public attention and speculation and may spark a new era in digitally monitoring in vivo glucose levels, but to be fair, the "Google contact lens" is expanding upon the previous efforts of many [35]. Google patents related to this contact lens (eye-mountable device) focus on two aspects: (i) the

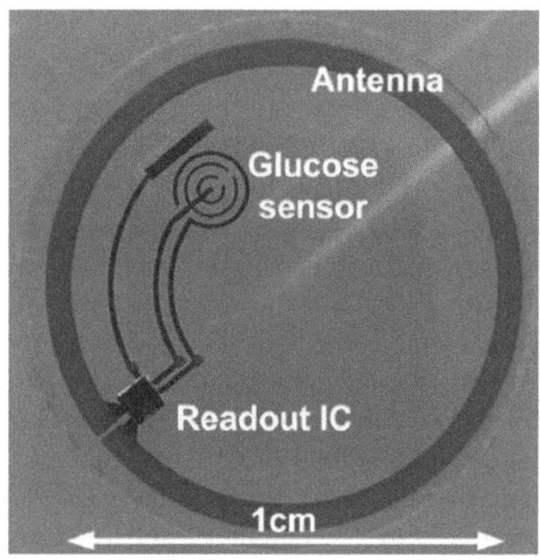

Fig. 4.3 Image of a contact lens with glucose sensor [30]

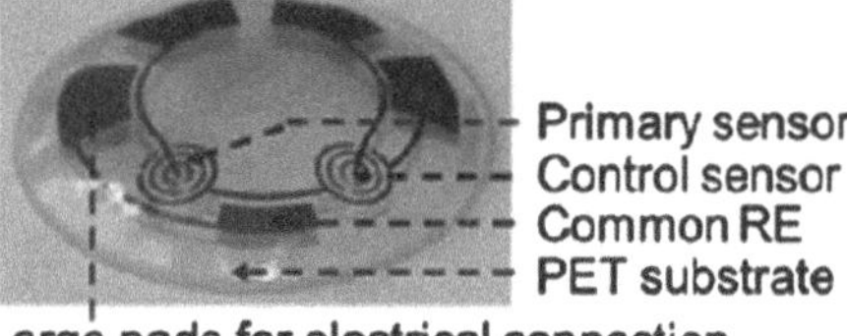

Fig. 4.4 Image of a contact lens with dual glucose sensors [33]

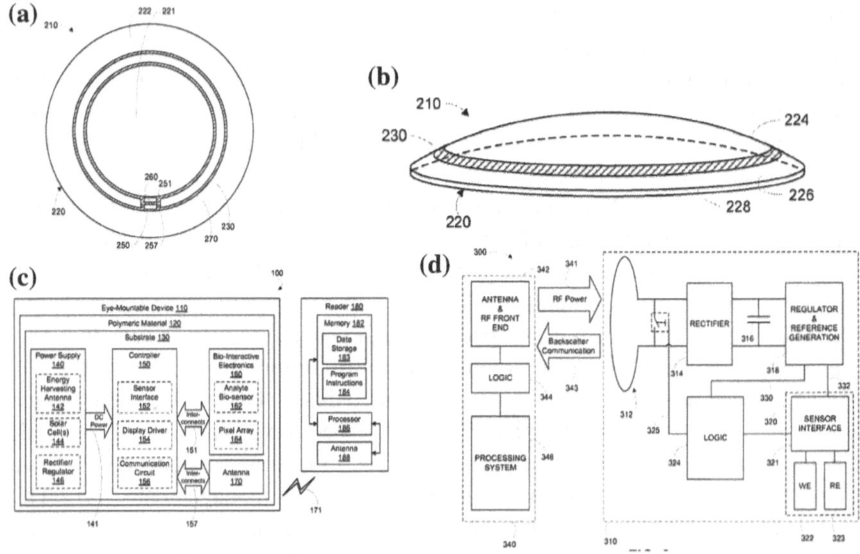

Fig. 4.5 An example of an eye-mountable device. **a**, **b** Bottom and side cross-sectional views, respectively. **c** A block diagram of an example system for an eye-mountable device with wireless communication to an external reader. **d** A functional block diagram of an example system for electrochemically measuring the concentration of analyte in tear fluid [38]

designs for components, i.e., controller, biointeractive electronics, and light source in the device and (ii) a device with dual power sources [36–43]. The transparent contact lens (diameter: approximately 1 cm; thickness: approximately 0.1–0.5 mm) has a concave surface that is compatible with eyelid motion, and a convex surface that includes a loop antenna, power supply, controller, and biointeractive electronics (all are embedded on conductive substrates) (Fig. 4.5) [38]. When the loop antenna receives incidental radio frequency radiation, the rectifier/regulator converts the RF signals into a proper DC voltage. Alternatively, solar cells could be used to supply the required energy for the whole system if the energy-harvesting antenna failed (dual power sources). After the controller accepts the energy, the controller (i) modulates an efficient voltage among working and reference electrodes to generate an aerometric current correlated with the different concentrations of analyte while the device exposing to tear sample, (ii) measures the aerometric current, and (iii) transmits the result after processing to readers (e.g., mobile phones) via antenna. The analyte biosensor comprises a working electrode (e.g., Ag/AgCl) that has at least one dimension <25 mm and a reference electrode that has an area at least five times greater than the area of the working electrode (Fig. 4.6) [38]. The working electrode is coated with GOD-catalyzed glucose and produces hydrogen peroxide. Following this, the working electrode accepts electrons and generates an amperometric current, while hydrogen peroxide is electro-oxidized at the working electrode. In the end, the current signal transmits to

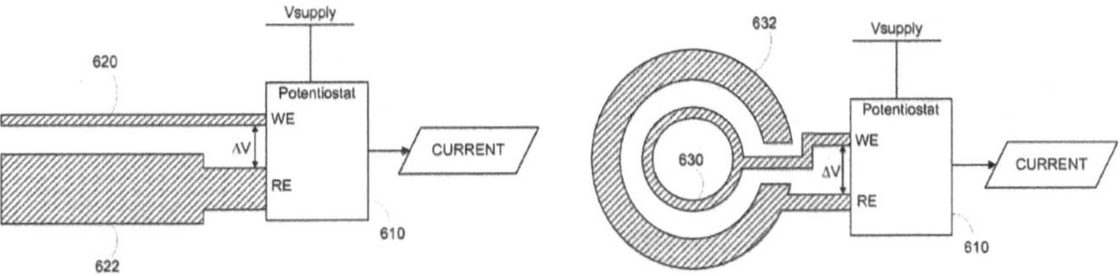

Fig. 4.6 Examples of arrangement for electrodes in electrochemical sensors [38]

readers for processing. In addition, the contact lens displays a pixel array (LED lights) that can be used to indicate insufficient power level to users. Accordingly, the measurement sensitivity of this device is highly correlated with the interaction between GOD and the sensors. Considering the detection sensitivity, there is a patent claimed that cationic polymers applied in a glucose sensor are able to improve detection sensitivity and increase enzyme stability at 37–65 °C [44]. With consideration still being given to contact lens wearing comfort, the idea of a hydrogel-based contact lens with large amounts of cavity appears feasible [45].

4.2 Design and Fabrication of the Contact Lens-Based Glucose Sensor

In this section, we will detail the fabrication of contact lens-based glucose sensors with a significant nod to the research of Brian Otis, whose research provided the blueprint for smart contact lens development [28–30, 33].

4.2.1 Glucose Sensor Design and Fabrication

Typically, a glucose sensor comprises three electrodes: (i) a working electrode (WE) coated with GOD that allows for glucose oxidation and accepts electrons when hydrogen peroxide is electro-oxidized; (ii) a counter electrode (CE) that acts as a current drain to form a current loop; and (iii) a reference electrode (RE) that supplies stable voltage potential to the whole system (Fig. 4.7) [30]. The WE and CE in the research reported on here were formed as concentric rings with widths of 50–75 μm, respectively [30, 46]. Both were equipped with a 50 μm pitch to reduce resistance and improve sensor sensitivity. The RE was designed as a rectangular bar (1.6 mm × 0.25 mm). To achieve high sensitivity and low inference, a modification of the glucose oxidase approach was employed. A thin piece of parafilm was used to cover the surface of the integrated contact lens but not the WE. The WE was treated with GOD (30 μL, 10 mg/mL), and then, the sensor was

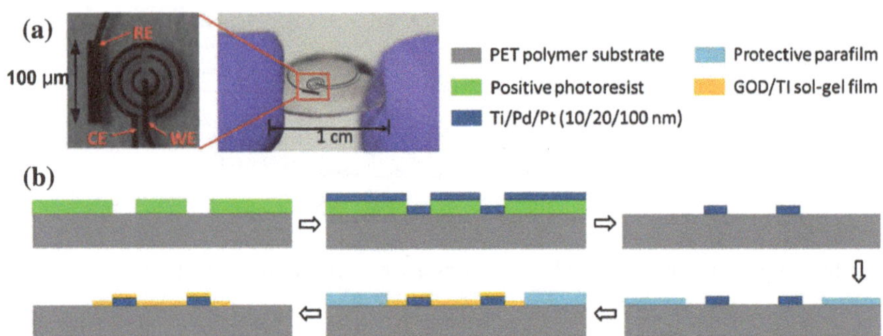

Fig. 4.7 Glucose sensor on a contact lens. **a** Design of electrode. **b** Fabrication of electrode [30]

suspended vertically above a titanium isopropoxide solution in a sealed dish for 6 h in order to create a GOD/titania solgel membrane. In the dual sensor design, a thin piece of parafilm was used to cover the surface of integrated contact lens, but not the control sensor. The control sensor was treated with DeGOD [GOD solution (10 mg/mL) at 70 °C for 3 h] (20 μL), and then, the sensor was suspended vertically above a titanium isopropoxide solution in a sealed dish for 6 h in order to create a GOD/titania solgel membrane. Next, the device was covered with a thin piece of parafilm except the primary sensor was not covered. The primary sensor went through the enzyme treatment process in the same manner as the control sensor steps, but GOD (20 μL, 10 mg/mL) was used [33].

4.2.2 LED (Red Light) Fabrication

The fabrication of LED lights is based on aluminum gallium arsenide (Fig. 4.8) [47]. In the research reported on here, an AlAs sacrificial layer was initially stacked on the top of a GaAs substrate [28]. Then, each individual LED layer was deposited onto the sacrificial layer. Afterward, etching and depositing Cr/Ni/Au were allowed to create n-region and p-region, respectively. In the end, hydrofluoric acid solution was used to release the LEDs from the wafer.

4.2.3 Antenna Design

A loop antenna has a radius of approximately 5 mm, a width of approximately 0.5 mm, and a thickness of approximately 5 μm. At that size, it does not obstruct the wearer's vision. The ICNIRP (International Committee on Non-ionizing Radiation Protection) regulates the general public head and trunk radio frequency exposure to 2 W kg^{-1} (up to 10 GHz), or a plane wave power

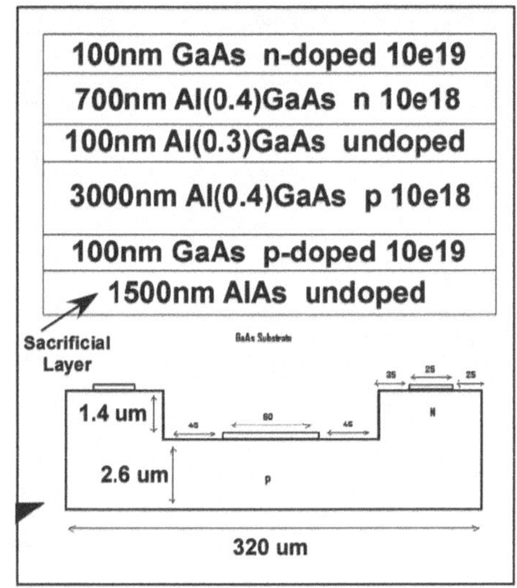

Fig. 4.8 Design of LED. **a** An example of fabrication of AlGaAs μ-LED. **b** Dimensions of μ-LED [47]

density of 10 W m^{-2} (2–300 GHz). The IEEE (Institute for Electrical and Electronics Engineers) regulates the exposure limitation of places at 2 W kg^{-1} (100 kHz–3 GHz)/10 W m^{-2} (2–100 GHz). In general, the RF operation is determined via (i) the amount of power received via a loop antenna and (ii) antenna-to-chip impedance matching. Considering these requirements, the resonant frequency of the antenna and chip system is approximately 980–990 MHz with a chip impedance of 7 Ω + 12 pF and an antenna inductance of 22 nH. The loop antenna receives power from a 1.0 W power source at a distance of 1.0 m and transmitted antenna gain of 1.76 dBi, while receiving power peaked at approximately 117 μW at approximately 2.0 GHz [29].

4.2.4 Wireless Readout Chip Architecture

The on-lens chip sensor readout system is designed to perform at low power (<5 μW) and low-current noise (<1 nA rms) in 0.36 mm^2 area. The IC is composed of a power management block, readout circuitry, wireless communication component, LED driver, and energy storage capacitor in a 0.36 mm^2 CMOS chip (Fig. 4.9) [30]. The production of energy in this chip is attributed to a RF-based energy-harvesting system receiving RF signal from the interrogator. In order to ameliorate the fluctuation of input energy (caused via varying strength of incident RF power as well as digital switching noise), the combination of an ultralow-power linear regulator, temperature-stable bandgap reference, and bias current generation provides stable DC bias and supply (+1.2 V) for the chip. In addition, a potentiate offers a stable potential (+400 mV) to initiate electro-oxidation

Fig. 4.9 Image of readout IC [30]

between the combination of WEs and CEs (the CE2, connected to a reference sensor, is designed to block the undesired interference) or the combination of WEs and REs. The measured current signal is amplified and converts to a specific frequency through the processing of an oscillator-based current-to-frequency converter. In the end, the system wirelessly communicates with the interrogator through RF backscatter.

4.2.5 Fabrication for the Integration of Radio and Sensor with Contact Lens

All construction is highly reliant on semiconductor-related techniques. The fabrication process flow is shown in Fig. 4.10 [33]. In order to preserve the fabrication process while considering biocompatibility issues, construction employed a polyethylene terephthalate (PET) substrate. A clean PET wafer was prepared (thickness ~0.1– 0.5 mm) (Fig. 4.10a). A thin AZ4620 film (6 μm) was spin-coated on the top of a PET wafer, and photolithography was used to pattern the area for a loop antenna. Then, three layers of metal film (Cr 20 nm, Ni 80 nm, and Au 350 nm) were evaporated onto the wafer, and the unnecessary AZ4620 was removed via acetone (lift-off), resulting in the antenna adhesion layer and electrical interconnects (Fig. 4.10b). To develop the sensor, the three layers of metal film (Ti 10 nm, Pd 20 nm, and Pt 100 nm) were deposited after pattering with AZ4620 photoresist (Fig. 4.10c). Subsequently, a thin SU-8 film (1.5 μm) with a defined pattern was coated onto the wafer to create the electrical insulation and solder wetting area (Fig. 4.10d). Finally, a layer of gold (40 nm) was evaporated onto the whole surface of wafer as a seed layer, and then, another layer of AZ4620

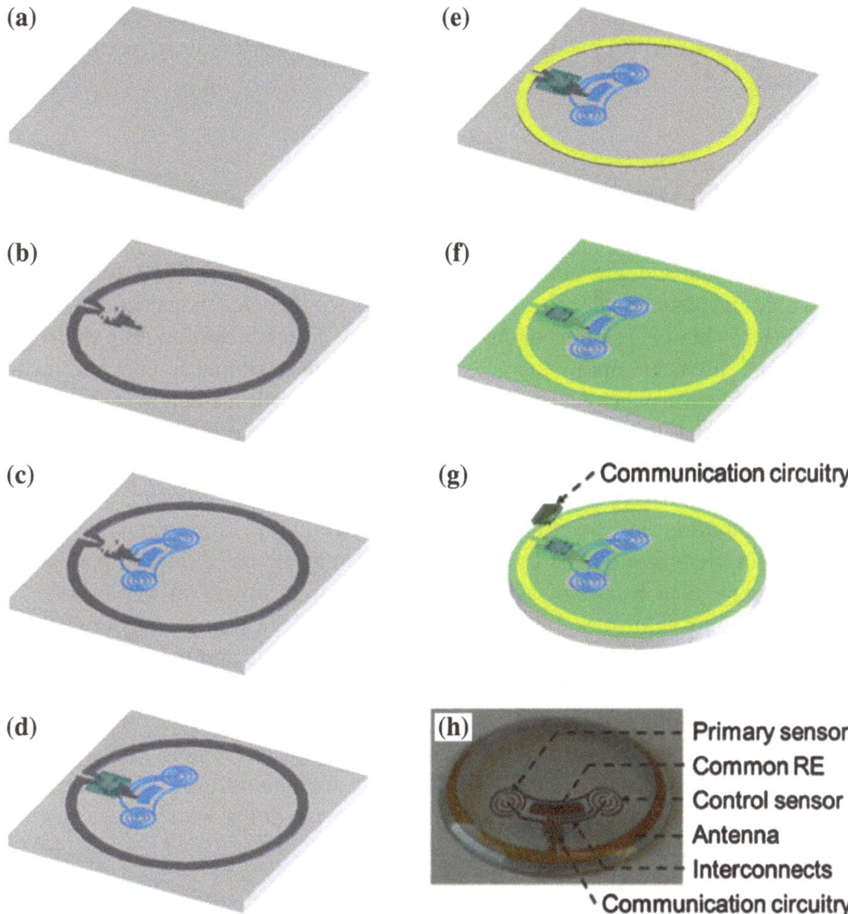

Fig. 4.10 Fabrication of a contact lens [33]. **a** Preparation of a clean PET substrate. **b** Fabrication of a Cr/Ni/Au layer and lift-off for building antenna adhesion layer and electrical interconnects. **c** Fabrication of a Ti/Pd/Pt layer and lift-off for establishing metal dual sensor. **d** A SU-8 layer covered on the surface of substrate for patterning the position of radio chip well. **e** Fabrication of a gold antenna. **f** Use of a SU-8 layer for protecting the metal structures. **g** Incorporation of a silicon chip on the device. **h** A contact lens with IC and sensor interface circuit

photoresist followed to act as a mask. As a result, a 5-μm-thick loop antenna with low electric resistance (ohmic) was produced through pulsed electroplating (1 ms on and 1 ms off) (Pur-A-Gold®401 formulation) (Fig. 4.10e). The photoresist was removed with acetone, and the seed layer was etched in 5 mL (vol:vol) deionized ddH_2O:Gold Etch TFA. Afterward, a thick layer SU-8 (25 μm) was coated onto the wafer to protect the metal structures and to open the spaces for the connection of dual sensors as well as a piece of silicon/LED chip (communication circuitry component) (Fig. 4.10f). Then, the silicon chip was incorporated onto the contact lens via molten-alloy-bridging self-assembly, and the integrated polymer was molded (200 °C, 15 s) and polished into the form of a contact lens (diameter approximately 1 cm) (Fig. 4.10g). Finally, the glucose sensors were modified with GOD (Sect. 2.1).

4.3 Potential Directions

In summary, the Google contact lens demonstrates the possibility for an eye-mountable device to monitor tear glucose levels for diabetes management, but its feasibility for human remains under investigation. These breakthroughs still forge new ground in the attempt to integrate the contact lens, a simple and popular medical aid, with electrochemical methodology for diagnostic detection. Google has also proposed that this device has the potential to reflect blood alcohol concentration via the use of an enzyme that can specifically oxidize alcohol so that future testing for drunk drivers might be relatively accessible and more precise compared with current urine and breath analysis approaches [48]. Moreover, the readout methods for capturing device processing results may be more diverse (computer, mobile phone, watch-based devices), which inspires growth in the frontier of wearable diagnostic devices [49]. Furthermore, such devices, e.g., a hydrogel contact lens, may be capable of providing ameliorative effects and not just diagnostic results as in the release of drugs to treat glaucoma [50]. We believe that integrating the functionalities of physical condition monitoring and delivering drugs via a contact lens is an achievement that could be realized in the near future. Instead of a contact lens with an exclusive functionality for monitoring or drug delivery, an advanced contact lens could automatically release drugs with precision when monitoring results determine a need.

References

1. Embuldeniya G, Veinot P, Bell E, Bell M, Nyhof-Young J, Sale JE, Britten N (2013) The experience and impact of chronic disease peer support interventions: a qualitative synthesis. Patient Educ Couns 1:3–12
2. Petersen PE, Ogawa H (2000) The global burden of periodontal disease: towards integration with chronic disease prevention and control. Periodontol 60:15–39
3. Vogeli C, Shields AE, Lee TA, Gibson TB, Marder WD, Weiss KB, Blumenthal D (2007) Multiple chronic conditions: prevalence, health consequences, and implications for quality, care management, and costs. J Gen Intern Med 22(Supply 3):391–395
4. http://www.cdc.gov/chronicdisease/resources/publications/aag/chronic.htm
5. Center on an Aging Society (2004) Disease management programs: improving health while reducing costs? Washington. Georgetown University, DC
6. American Diabetes Association (2008) Economic costs of diabetes in the U.S. in 2007. Diabetes Care 31:596–615
7. Egede LE, Gebregziabher M, Zhao Y, Dismuke CE, Walker RJ, Hunt KJ, Axon RN (2014) Impact of mental health visits on healthcare cost in patients with diabetes and comorbid mental health disorders. PLoS ONE 9:e103804

8. Fisher L, Skaff MM, Mullan JT, Arean P, Glasgow R, Masharani U (2008) A longitudinal study of affective and anxiety disorders, depressive affect and diabetes distress in adults with type 2 diabetes. Diabet Med 25:1096–1101
9. Yo EH, Lee SY (2010) Glucose biosensors: an overview of use in clinical practice. Sensors 10:4558–4576
10. Hiratsuka A, Fujisawa K, Muguruma H (2008) Amperometric biosensor based on glucose dehydrogenase and plasma-polymerized thin films. Anal Sci 24:483–486
11. American Diabetes Association (1996) American Diabetes Association: clinical practice recommendations 1996. Diabetes Care 19:S1–S118
12. Solnica B, Naskalski JW (2007) Quality control of self-monitoring of blood glucose: why and how? J Diabetes Sci Technol 1:164–168
13. Bandodkar AJ, Wang J (2014) Non-invasive wearable electrochemical sensors: a review. Trends Biotechnol 32:363–371
14. Martinez AW, Phillips ST, Carrilho E, Thomas SW, Sindi H, Whitesides GM (2008) Simple telemedicine for developing regions: camera phones and paper-based microfluidic devices for real-time, off-site diagnosis. Anal Chem 80:3699–3707
15. Lee DS, Jeon BG, Ihm C, Park JK, Jung MY (2011) A simple and smart telemedicine device for developing regions: a pocket-sized colorimetric reader. Lab Chip 11:120–126
16. Mudanyali O, Dimitrov S, Sikora U, Padmanabhan S, Navruz I, Ozcan A (2012) Integrated rapid-diagnostic-test reader platform on a cellphone. Lab Chip 12:2678–2686
17. Wang S, Zhao X, Khimji I, Akbas R, Qiu W, Edwards D, Cramer DW, Ye B, Demirci U (2011) Integration of cell phone imaging with microchip ELISA to detect ovarian cancer HE4 biomarker in urine at the point-of-care. Lab Chip 11:3411–3418
18. Shen L, Hagen JA, Papautsky I (2012) Point-of-care colorimetric detection with a smartphone. Lab Chip 12:4240–4243
19. Matlani P, Londh ND (2013) A cloud computing based telemedicine service. Point-of-Care Healthcare Technologies (PHT), 2013 IEEE, pp 326–330
20. Schazmann B, Morris D, Slater C, Beirne S, Fay C, Reuveny R, Moynac N, Diamond D (2010) A wearable electrochemical sensor for the real-time measurement of sweat sodium concentration. Anal Methods 2:342–348
21. Guinovart T, Parrilla M, Crespo GA, Rius FX, Andrade FJ (2013) Potentiometric sensors using cotton yarns, carbon nanotubes and polymeric membranes. Analyst 138:5208–5215
22. Jia W, Bandodkar AJ, Valdés-Ramírez G, Windmiller JR, Yang Z, Ramírez J, Chan G, Wang J (2013) Electrochemical tattoo biosensors for real-time noninvasive lactate monitoring in human perspiration. Anal Chem 85:6553–6560
23. Mannoor MS, Tao H, Clayton JD, Sengupta A, Kaplan DL, Naik RR, Verma N, Omenetto FG, McAlpine MC (2012) Graphene-based wireless bacteria detection on tooth enamel. Nat Commun 3:763
24. Yan Q, Peng B, Su G, Cohan BE, Major TC, Meyerhoff ME (2011) Measurement of tear glucose levels with amperometric glucose biosensor/capillary tube configuration. Anal Chem 83:8341–8346
25. http://www.sensimed.ch/en/company/about-us.html
26. March WF, Mueller A, Herbrechtsmeier P (2004) Clinical trial of a noninvasive contact lens glucose sensor. Diabetes Technol Ther 6:782–789
27. Chu M, Shirai T, Takahashi D, Arakawa T, Kudo H, Sano K, Sawada S, Yano K, Iwasaki Y, Akiyoshi K, Mochizuki M, Mitsubayashi K (2011) Biomedical soft contact-lens sensor for in situ ocular biomonitoring of tear contents. Biomed Microdevices 13:603–611
28. Pandey J, Liao Y, Lingley AR, Mirjalili R, Parviz BA, Otis BP (2010) A fully integrated RF powered contact lens with a single element display. IEEE Trans Biomed Circuits Syst 4:454–461
29. Lingley AR, Ali M, Liao Y, Mirjalili R, Klonner M, Sopanen M, Suihkonen S, Shen T, Otis BP, Lipsanen H, Parviz BA (2011) A single-pixel wireless contact lens display. J Micromech Microeng 21:125014

30. Liao Y, Yao H, Lingley A, Parviz BA, Otis BP (2012) A 3 μW CMOS glucose sensor for wireless contact-lens tear glucose monitoring. J Solid-State Circuits 47:335–344
31. Bandodkar AJ, Wang J (2014) Non-invasive wearable electrochemical sensors: a review. Trends Biotechnol 32:363–371
32. Berman ER (1991) Biochemistry of the eye. Plenum, New York, pp 68–76
33. Yao H, Liao Y, Lingley AR, Afanasiev A, Lähdesmäki I, Otis BP, Parviz BA (2012) A contact lens with integrated telecommunication circuit and sensors for wireless and continuous tear glucose monitoring. J Micromech Microeng 22:075007
34. http://www.forbes.com/sites/leoking/2014/07/15/google-smart-contact-lens-focuses-on-healthcare-billions/
35. Otis B, Liao YT, Amirparviz B, Yao H (2012) Wireless powered contact lens with glucose sensor. US Pat 20120245444:A1
36. Liu Z, Otis B (2014) In-vitro calibration of an ophthalmic analyte sensor. US Pat 20140107447:A1
37. Liu Z, Otis B (2014) In-vitro calibration of an ophthalmic analyte sensor. US Pat 20140107448:A1
38. Liu Z (2014) Microelectrodes in an ophthalmic electrochemical sensor. US Pat 20140107444:A1
39. Liu Z (2014) Contact lenses having two-electrode electrochemical sensors. US Pat 20140194713:A1
40. Liu Z (2014) Contact lenses having two-electrode electrochemical sensors. US Pat 20140190839:A1
41. Biederman WJ, Otis B (2014) Devices and methods for a contact lens with an inward facing light source. US Pat 8764185:B1
42. Nelson P, Liu Z, Otis B (2014) Facilitation of temperature compensation for contact lens sensors and temperature sensing. US Pat 20140085600:A1
43. Biederman WJ, Pletcher N, Nelson A, Yeager D (2014) Device with dual power sources. US Pat 8742623:B1
44. Feldman BJ, Ouyang T, Liu Z (2014) Cationic polymer based wired enzyme formulations for use in analyte sensors. US Pat 20140141487:A1
45. Liu Z, Etzkorn J (2014) In-situ tear sample collection and testing using a contact lens. US Pat 20140194706:A1
46. Yao H, Shum AJ, Cowan M, Lähdesmäki I, Parviz BA (2011) A contact lens with embedded sensor for monitoring tear glucose level. Biosens Bioelectron 26:3290–3296
47. Saeedi E, Kim SS, Parviz BA (2007) Self-assembled inorganic micro-display on plastic. In: Proceedings of IEEE 20th international conference on MEMS, pp 755–758
48. Liu Z, Amirparviz B (2014) Sensor. US Pat 20140206966:A1
49. https://www.apple.com/watch/technology/
50. Ciolino JB, Stefanescu CF, Ross AE, Salvador-Culla B, Cortez P, Ford EM, Wymbs KA, Sprague SL, Mascoop DR, Rudina SS, Trauger SA, Cade F, Kohane DS (2014) In vivo performance of a drug-eluting contact lens to treat glaucoma for a month. Biomaterials 35:432–439

Zeitfracht Medien GmbH
Ferdinand-Jühlke-Straße 7
99095 Erfurt, Deutschland
produktsicherheit@kolibri360.de